ADVANCES IN PLASMA PHYSICS
Thomas H. Stix Symposium

Thomas H. Stix

AIP
CONFERENCE
PROCEEDINGS 314

ADVANCES IN
PLASMA PHYSICS
Thomas H. Stix Symposium

PRINCETON, NJ

EDITOR: NATHANIEL J. FISCH
PRINCETON UNIVERSITY

American Institute of Physics   New York

L.C. Catalog Card No. 94-72721
ISBN 1-56396-372-8
DOE CONF-9205408

Printed in the United States of America.

Contents*

*NOTE: Underline denotes presenting author.

PREFACE

On May 4 and 5, 1992, a symposium was held in Princeton, New Jersey in honor of Professor Thomas H. Stix, who had stepped aside as Director of the Program in Plasma Physics at Princeton University. Part of the scientific program of the symposium included invited lectures. This book is a compendium of articles based on most of those lectures.

It is worthwhile to review briefly the career that led to this fest: After serving in the United States Army from 1942 to 1945, and after receiving his B.S. from the California Institute of Technology in 1948, Thomas Stix received his Ph.D. from Princeton in 1953, and joined Project Matterhorn, a small project started by Lyman Spitzer on Princeton's Forrestal Campus. The modest project aimed very high—to harness fusion energy for peacetime use.

Project Matterhorn grew fast, and in 1961, its name was changed to the Princeton Plasma Physics Laboratory. Tom Stix coheaded the experimental division from 1957 to 1978, after which he headed the Basic Plasma Physics Group. Also, in 1962, he was appointed Professor of Astrophysical Sciences at Princeton University and given primary responsibility for the then fledgling graduate program in plasma physics; he was later appointed Associate Director for Academic Affairs at PPPL, and for many years he was Associate Chairman of the Department of Astrophysical Sciences.

When one thinks of Tom, two things spring to mind: plasma waves and graduate education; conversely, it is impossible to think of plasma waves or graduate education without bringing to mind the name of Tom Stix. In fact, evident throughout the symposium was the dual career that Tom Stix pursued as an educator and as a researcher.

One reason that these things spring to mind is the book that everybody uses, Tom's book on plasma waves. In the 1958 Atoms for Peace Conference in Geneva, Tom revolutionized research in plasma waves by showing how plasma waves could be utilized to heat a plasma. Growing out of his research efforts at finding some way of igniting plasma by this supplementary heating, and out of his efforts at teaching students entering the new, exciting, and recently declassified field of thermonuclear fusion, Tom followed up on the Geneva Conference by publishing, in 1962, the classic text on plasma waves, *The Theory of Plasma Waves*. In educating a generation and more of plasma physicists, with 6000 copies in print, no book has had greater impact.

His achievement in education is evidenced also by the Program in Plasma Physics that he built. There are now over 160 graduates of this Program, many of whom are now leaders in the field themselves, and who had a reunion of sorts at the "Stix Symposium." It was fitting and interesting, therefore, that in addition to the invited lectures, the scientific program consisted of many short presentations of current research by these graduates. Seeing at once the scientific accomplishments of these graduates left all of us with an enriched perspective of the monumental educational legacy left by Tom.

Important as the educational legacy is, it is Tom's ideas in plasma physics, however, that ultimately leave the most personalized and imaginative mark on the field. His research fueled the rapid development of the field of controlled thermonuclear fusion, particularly in the important areas of plasma confinement and heating. His pioneering research achievements include the conceiving and the experimental demonstration of ion cyclotron heating of early stellarators, the first divertor experiments, and the first U.S. tokamak geometry experiments in 1957. In addition, he is known for his theoretical work on plasma wave mode conversion, on neutral beam heating, and, with Alexander Rechester, on transport in stochastic magnetic fields. Wide attention has been given to his recent suggestions on using infrared lasers to destroy atmospheric chlorofluorocarbons (CFCs). His 1975 paper on fast wave heating and quasilinear diffusion of a plasma is one of the most often cited papers ever published in Nuclear Fusion.

The book jacket shows a schematic drawing of the B-65 stellarator. The B-64 (the odd numerical nomenclature is derived imaginatively from the figure-8 geometry but with "squared" corners) and B-65 stellarators were designed and built by Tom Stix in the late 1950s, with assistance from Richard Palladino. A source of pride to these builders must be the improvisations made in the early days of fusion devices; the epoxy on the B-64 stellarator was cured using a Sears kitchen range. The B-65 stellarator, however, featured a radio frequency coil antenna, in which sections of coil were alternately wound around the device clockwise and counter-clockwise. Later known as a Stix coil, this structure efficiently coupled radio frequency waves at ion cyclotron frequencies into the plasma.

It has been recognized that some form of auxiliary heating would be necessary to bring a fusion reactor to the temperatures at which controlled thermonuclear ignition would occur (about 50 million degrees centigrade). Magnetic pumping by low frequency waves had been tried with only marginal success. However, the relatively high frequency waves at the ion cyclotron frequency excited by the Stix coil showed impressive results. The B-66 device, managed by Bill Hooke, continued to explore ion cyclotron resonance heating. At present, ion cyclotron resonance heating (ICRH) and other radio frequency heating schemes are envisioned as the most practical means of igniting plasma to ignition, and this continues to be an area of intense research 35 years after the B-64/65/66 devices.

Incidentally, on a personal note, my own scientific debt is owed to Tom, because many of the same radio frequency waves that he explored for heating a plasma to ignition later found a related use in driving the toroidal currents necessary for confinement in tokamaks. The use of these waves for current drive, as well as for heating, makes feasible a steady state tokamak. By careful attention to Tom's writings on the problem of plasma heating, I, together with many other researchers, could build on his thoughts in this other important area.

Tom Stix was the recipient of numerous awards, including predoctoral and postdoctoral fellowships, a Guggenheim Fellowship, and in 1980 he was the sixth recipient of the highest honor that the American Physical Society Division of Plasma Physics confers, the Maxwell Prize. The Maxwell Prize citation reads:

For his contributions to the development and formalization of the theory of wave propagation in plasmas and for his pioneering research on radio frequency plasma heating. His work played the guiding role in the understanding of space plasmas and in the development of advanced plasma heating methods for controlled fusion devices.

In addition, Tom served on numerous editorial boards, national committees, and panels, and he chaired the Division of Plasma Physics. In recent years, he has been particularly active in the area of human rights. His great success as a teacher and educator was recognized in 1992 by Princeton University in its awarding him its first University Award for Distinguished Teaching.

The present compendium of papers, based on the invited lectures, reflects aspects of Tom's career in many ways. The discussion of ICRF heating, by Joel Hosea, gives the perspective in which one can understand the profound influence that Tom Stix had on the field of heating and diagnostics. The paper by Miklos Porkolab gives a sense of current experimentation in this area. One may note that several of the speakers graduated the Program in Plasma Physics: Herbert Berk '64, Ronald Davidson '66, Charles Kennel '64, William Kruer '69, Masa Ono '78, Steven Orszag '66, and Alfred Wong '63. Shih-Tung Tsai '69, who worked closely with Tom during the period of his graduate studies, contributed a paper to this compendium, although he was unable to attend the symposium itself.

Considering all that he has done, it is not hard to imagine that the Stix Symposium attracted a huge audience attended by a festive atmosphere. After all, so many of us in the field had been touched by his teaching, his research, and his scientific leadership. It would be impossible to capture quite the atmosphere, but I will recall just a few things: the banquet was suitably reverent and irreverent with superb mastery of the ceremonies by Professor Martin Kruskal; the humor reached its peak in the last speech, Tom's rejoinder (in which he thanked the speakers for what they did not say); and, most importantly, a great warmth of appreciation for Tom's accomplishments exuded.

Something else was also evident at this banquet, and this relates to a different preface. In 1992, a second edition of "the book" was published under the title *Waves in Plasmas*. In addition to other new material, there is a nice preface to that book in which Tom Stix again acknowledges certain contributions of his wife Hazel, and that he always took her at her word. It was evident throughout the symposium that Tom was taken at his word, for great appreciation also exuded for Hazel, who accompanies Tom in his many adventures and travels in the world of physics, and who is appreciated for her gracious hosting of departmental parties and for her concern for Tom's colleagues and students.

In arranging the scientific program, my thanks go to the Scientific Steering Committee consisting of Professors Ronald Davidson, Miklos Porkolab, and Alfred Wong. In addition, Ms. Barbara Sarfaty is thanked by all of us for critical help in making the symposium a success.

Nathaniel J. Fisch
Princeton, New Jersey
April 1994

Advances in Plasma Physics
Thomas H. Stix Symposium
May 4 and 5, 1992

Executive Committee

Ronald C. Davidson

Princeton University, Plasma Physics Laboratory

Nathaniel J. Fisch (Chairman)

Princeton University, Plasma Physics Laboratory

Miklos Porkolab

Plasma Fusion Center, Massachusetts Institute of Technology

Alfred Y. Wong

University of California, Los Angeles

Symposium Coordinator

Barbara Sarfaty

Princeton University, Plasma Physics Laboratory

Attendees

Alumni Attendees

Attendee List

Jay Albert '85, Air Force Geophysics Laboratory

W. Arunsalam, PPPL

Peter Beiersdorfer '88, LLNL

Herbert Berk '64, IFS, University of Texas at Austin

Amitava Bhattacharjee '80, Columbia University

Stephen Bodner '65, NRL

Rejean Boivin '91, MIT

David Book '64

Bas Braams, CIMS, NYU

Joseph Cecchi, PPPL

Morrell Chance, PPPL

Liu Chen, PPPL

Frank Cheng, PPPL

Sam Cohen, PPPL

Steve Cowley '85, PPPL

Doug Darrow '88, PPPL

Ronald Davidson '66, PPPL

John Dawson, UCLA

C. Richard DeVore '86, NRL

Daniel Dubin '84, UCSD

Richard Ellis '70, University of MD

Nathaniel Fisch, PPPL

Harold Furth, PPPL

Robert Goldston '77, PPPL

Melvin Gottlieb, PPPL

Taik-Soo Hahm '84, PPPL

Gregory Hammett '86, PPPL

Adilnawaz Hassam '78, University of MD

Richard Hawryluk, PPPL

Rush Holt, PPPL

Joel Hosea, PPPL

Hulbert Hsuan '67, PPPL

David Ignat, PPPL

Stephen Jardin '76, PPPL

Forest Jobes, PPPL

John Johnson, PPPL

Robert Kaita, PPPL

Charles Karney, PPPL

Charles Kennell '64, UCLA

Arnold Kritz, Lehigh College

John Krommes '75, PPPL

William Kruer '69, LLNL

Martin Kruskal, Rutgers University

Russell Kulsrud, PPPL/Princeton University

Wei-le Lee, PPPL

Christof Litwin, University of Wisconsin

Richard Marchand '79, INRS

Michael Mauel, Columbia University

Dale Meade, PPPL

Howard Milchberg '85, University of MD

Harry Mynick, PPPL

Gerald Navratil, Columbia University

Masayuki Ono '78, PPPL

Joseph Orens '74, Berkeley Research Association

Steven Orszag '66, Princeton University

Richard Palladino, PPPL

Francis Perkins, PPPL

Miklos Porkolab, MIT

Douglass Post, PPPL

Jean-Marcel Rax, PPPL

Alexander Rechester, Institute of Nonlinear Science Applications

Martha Redi, PPPL

Dick Rossi, PPPL

Paul Rutherford, PPPL

Walter Sadowski, DoE

Ned Sauthoff '75, PPPL

John Schmidt, PPPL

James Sinnis, PPPL

Fred Skiff '84, University of MD

Lyman Spitzer, Princeton University

Jim Stevens, PPPL

Szymon Suckewer, PPPL/Princeton University

William Tang, PPPL

Ernest Valeo '71, PPPL

Schwick von Goeler, PPPL

Roscoe White, PPPL

Randy Wilson, PPPL

Alfred Wong '63, UCLA

King-Lap Wong, PPPL
Glen Wurden '82, PPPL

Masaaki Yamada, PPPL
Shoichi Yoshikawa, PPPL

Graduate Students
Princeton University
Department of Astrophysical Sciences
Program in Plasma Physics

Michael Beer
Gordon Chiu
Wonho Choe
David Coster
Julian Cummings
William Dorland
Genze Hu
Yong-Seok Hwang
Theodore Jones
Karl Krushelnick
Zhihong Lin
Ernest Lo

Hui Long
Alexander MacAulay
Q. Peter Liu
Sherrie Preische
Qian Qian
John Reynders
Stephen Smith
George Vetoulis
Keith Voss
Yanlin Wu
Yi Zhao

Symposium Program

Monday, May 4, 1992

8:30 a.m.	Welcome, Nathaniel J. Fisch
8:40	Opening Remarks, Lyman Spitzer

Chairman: Melvin B. Gottlieb

8:50 a.m.	Ronald C. Davidson (PPPL) *"Advances in Nonneutral Plasmas"*
9:25	William L. Kruer (LLNL) *"Ultra-Intense Laser Plasma Interactions"*
10:00	Coffee Break

Chairman: Harold Furth

10:20 a.m.	Steven A. Orszag (Princeton University) *"Random Rayleigh-Taylor Instability"*
10:55	Alumni
11:25	Martin D. Kruskal (Rutgers University) *"Asymptotics Beyond All Orders"*
12:00 p.m.	Lunch

Chairman: Russell M. Kulsrud

1:30 p.m.	John M. Dawson (UCLA) *"Possibilities for Steady-State Tokamaks"*
2:05	Alumni
2:30	Alfred Y. Wong (UCLA) *"Mitigation of Ozone Depletion by Charging the Atmosphere"*
3:05	Coffee Break

Chairman: William Nevins

3:30 p.m.	Alumni
3:50	Miklos Porkolab (MIT) *"Electron Heating and Current Drive in Tokamaks by Fast Magnetosonic Waves"*

Banquet—Scanticon Hotel

6:30 p.m.	Cocktails
7:30	Dinner

Tuesday, May 5, 1992

Chairman: Szymon Suckewer

9:00 a.m.	Alexander Rechester (Institute of Nonlinear Science Applications) *"Symbolic Signal Processing and Pattern Recognition "*
9:35	Joel C. Hosea (PPPL) *"Foundations of ICRF Heating—A Historical Perspective "*
10:10	Herbert L. Berk (IFS, University of Texas) *"Pulsations and Explosions with Weak Beam Instabilities "*
10:45	Coffee Break

Chairman: Christof Litwin

11:05 a.m.	Masayuki Ono (PPPL) *"Basic Toroidal Plasma Experiments"*
11:25	Alumni
12:05 p.m.	Charles F. Kennel (UCLA) *"New Results in the Theory of Intermediate Shocks"*
12:40	Closing Remarks
1:00	Lunch

THERMAL EQUILIBRIUM PROPERTIES OF NONNEUTRAL PLASMA IN THE WEAK COUPLING APPROXIMATION

Ronald C. Davidson and Steven M. Lund[†]

Princeton Plasma Physics Laboratory

Princeton University, Princeton, NJ 08543

ABSTRACT

Thermal equilibrium properties are calculated for a cylindrical, pure electron plasma confined radially by a uniform axial magnetic field $B_0 \mathbf{e}_z$. In the weak coupling approximation $(e^2/\hat{n}_e^{-1/3} \ll k_B T_e)$, the one-particle thermal equilibrium distribution function is $f_{eq}(H, P_\theta) = const. \times \exp\{-(H - \omega_r P_\theta)/k_B T_e\}$, where H is the energy, P_θ is the canonical angular momentum, T_e is the temperature, and ω_r is the equilibrium angular rotation velocity of an electron fluid element. The self-consistent equilibrium density profile $n_{eq}(r) = \int d^3p\, f_{eq}$ is characterized over a wide range of values of the on-axis electron density $(\hat{n}_e)$, electron temperature (T_e), and confining field strength (B_0). Closed analytical expressions are derived for the mean-square radius $\langle r^2 \rangle_{eq}$ of the plasma column and the angular rotation velocity ω_r, expressed in terms of the nonlinear conservation constraints, $N_e = \int d^2x \int d^3p\, f_e(\mathbf{x}, \mathbf{p}, t) = const.$ and $\langle P_\theta \rangle = N_e^{-1} \int d^2x \int d^3p\, P_\theta f_e(\mathbf{x}, \mathbf{p}, t) = const.$, which correspond to the total number of electrons per unit axial length and the total canonical angular momentum per unit axial length, respectively. Finally, an exact power-series solution is derived for the thermal equilibrium density profile $n_{eq}(r)$ as a function of the radial distance r from the axis of the plasma column.

[†]Permanent address, Lawrence Livermore National Laboratory, Livermore, CA 94550

I. INTRODUCTION

Recent experimental and theoretical investigations of the fundamental properties of nonneutral plasmas[1,2] have included phenomena as diverse as coherent nonlinear structures,[3-7] chaotic and stochastic effects,[8,9] phase transitions in strongly-coupled two- and three-dimensional nonneutral plasmas,[10-13] excitation of collective oscillations and instabilities,[14,15] transport properties[16-18] and approach to thermal equilibrium,[17-20] astrophysical studies of large-scale isolated nonneutral plasma regions in the magnetospheres of rotating neutron stars,[21] and the development of positron and antiproton storage traps,[22-24] to mention a few examples. One-component nonneutral plasmas are found to exhibit remarkable stability and confinement properties,[1,2] due in no small measure to the powerful influence of the nonlinear conservation constraints which are applicable in macroscopic,[25] kinetic,[19,20] and microscopic[26] models of such systems. Furthermore, a paradigm has been developed which provides an algorithm[1,19,20] for relating the waves and instabilities observed in uniform, nonneutral plasma to the analogous collective phenomena in electrically neutral plasma.[27] In addition, it has been shown that the cold-fluid guiding-center model describing the two-dimensional nonlinear evolution of low-density, strongly magnetized nonneutral plasma is isomorphic to the Euler equations governing the flow of a constant-density inviscid fluid in two dimensions.[3]

In view of the extensive effort to investigate fundamental processes in nonneutral plasmas, particularly in the weak-coupling regime, it is of considerable importance to understand the basic properties of nonneutral plasma in thermal equilibrium. The basic equations[1] that describe the thermal equilibrium properties of a nonneutral plasma column confined radially by a uniform magnetic field $B_0 e_z$ have been known for some time and analyzed numerically.[1,28] In addition, collisional transport models describing the approach to thermal equilibrium have been developed in the low-density regime,[16,17] and detailed experimental investigations[18] have been carried out showing relaxation to the expected thermal equilibrium density profile $n_{eq}(r)$. Nonetheless, given the strong influence of equilibrium self-electric-field effects and the highly nonlinear nature of Poisson's equation in thermal equilibrium, there is a paucity of analytical results describing simple properties of the equilibrium. The primary purpose of this paper is to determine certain key properties of the thermal equilibrium analytically, such as the mean-square radius $\langle r^2 \rangle_{eq}$, and to relate these properties to the nonlinear conservation constraints corresponding to the total number of electrons per unit axial length N_e, and the total canonical angular momentum per unit axial length $\langle P_\theta \rangle$. In addition, an exact power-series solution is derived for the thermal equilibrium density profile $n_{eq}(r)$ as a function of the radial distance r from the axis of the plasma column.

II. THERMAL EQUILIBRIUM MODEL

In the present analysis, we investigate several properties of a cylindrical nonneutral plasma in thermal equilibrium that can be calculated in closed analytical form. The pure electron plasma is confined radially by a uniform axial magnetic field $B_0 \mathbf{e}_z$, where $B_0 = const.$ is assumed (negligible diamagnetism), and a perfectly conducting cylindrical wall is located at radius $r = r_w$, where $r_w \gg r_p$. Here, r_p is the characteristic radius of the plasma. For an infinitely long plasma column, the one-particle thermal equilibrium distribution function $f_{eq}(\mathbf{x}, \mathbf{p})$ can be expressed as[1,19,20]

$$f_{eq}(H, P_\theta) = \frac{\hat{n}_e}{(2\pi m k_B T_e)^{3/2}} \exp\left\{ -\frac{H - \omega_r P_\theta}{k_B T_e} \right\} . \tag{1}$$

However complex the initial distribution $f_e(\mathbf{x}, \mathbf{p}, t = 0)$ and the concomitant nonlinear evolution of $f_e(\mathbf{x}, \mathbf{p}, t)$ through collective and collisional processes, in the weak coupling approximation ($e^2/\hat{n}_e^{-1/3} \ll k_B T_e$) the one-particle distribution approaches the thermal equilibrium form in Eq. (1) after several binary collision times.[16-18] In Eq. (1), $\hat{n}_e$ is a positive constant, T_e is the electron temperature, k_B is Boltzmann's constant, $H = (p_r^2 + p_\theta^2 + p_z^2)/2m - e\phi(r)$ is the particle energy, $\phi(r)$ is the equilibrium electrostatic potential, $P_\theta = r(p_\theta - mr\omega_c/2)$ is the canonical angular momentum, m and $-e$ are the electron mass and charge, respectively, $\omega_c = eB_0/mc$ is the electron cyclotron frequency in the applied magnetic field $B_0 \mathbf{e}_z$, and c is the speed of light in vacuo. Here, we have introduced cylindrical polar coordinates (r, θ, z), where $r = (x^2 + y^2)^{1/2}$ is the radial distance from the axis of the plasma column. In addition, the rotational frequency $\omega_r = const.$ occurring in the exponent in Eq. (1) can be identified with the angular rotation velocity of a plasma fluid element calculated from Eq. (1), i.e., $\omega_r r = (\int d^3 p\, f_{eq} p_\theta/m)/(\int d^3 p\, f_{eq})$. Note from Eq. (1) that the present analysis assumes zero average axial velocity of the plasma electrons, i.e., $0 = \int d^3 p\, f_{eq} p_z/m$. [The generalization of Eq. (1) to the case where the average axial velocity of the plasma electrons is $V_z = const.$ is obtained by making the replacement $H - \omega_r P_\theta \rightarrow H - \omega_r P_\theta - V_z p_z$ in Eq. (1).] Some straightforward algebra shows that the equilibrium electron density profile $n_{eq}(r) = \int d^3 p\, f_{eq}(H, P_\theta)$ calculated from Eq. (1) can be expressed as

$$n_{eq}(r) = \hat{n}_e \exp\left\{ -\frac{m}{2k_B T_e} \left[r^2(\omega_r \omega_c - \omega_r^2) - \frac{2e}{m}\phi(r) \right] \right\} . \tag{2}$$

Without loss in generality we take $r = 0$ to correspond to the zero of electrostatic potential, i.e., $\phi(r = 0) = 0$, and the conducting wall at radius $r = r_w$ is at constant potential $\phi(r = r_w) = V = const.$ that is determined from Poisson's equation. Therefore, from Eq. (2), $\hat{n}_e = const.$ can be identified with the on-axis value of equilibrium electron density, i.e., $\hat{n}_e \equiv n_{eq}(r = 0)$. Of course, the equilibrium electrostatic potential $\phi(r)$ occurring in Eqs. (1) and (2) is determined self-consistently from Poisson's equation $r^{-1}(\partial/\partial r)[r\partial\phi/\partial r] = 4\pi e n_{eq} = 4\pi e \int d^3p\, f_{eq}$, which can be expressed as

$$\frac{1}{r}\frac{\partial}{\partial r} r \frac{\partial}{\partial r}\phi(r) = 4\pi e \hat{n}_e \exp\left\{-\frac{m}{2k_B T_e}\left[r^2(\omega_r\omega_c - \omega_r^2) - \frac{2e}{m}\phi(r)\right]\right\} . \qquad (3)$$

Note that Eq. (3) is a nonlinear differential equation for the equilibrium electrostatic potential $\phi(r)$ which can be solved numerically. Later in this article, we present an exact power-series solution for $n_{eq}(r)$ obtained from Eqs. (2) and (3).

For radially confined equilibria with $n_{eq}(r \to \infty) = 0$, it is necessary that $\omega_r\omega_c - \omega_r^2 \geq \hat{\omega}_p^2/2$, where $\hat{\omega}_p^2 = 4\pi\hat{n}_e e^2/m$ is the on-axis plasma frequency-squared. Denoting the positive dimensionless quantity ϵ by

$$\frac{\omega_r\omega_c - \omega_r^2}{\hat{\omega}_p^2/2} = 1 + \epsilon , \qquad (4)$$

it can be shown[1] from Eqs. (2) and (3), whenever $\epsilon \ll 1$, that the equilibrium density profile $n_{eq}(r)$ is approximately uniform in the column interior [$n_{eq}(r) \simeq \hat{n}_e$ for $0 \leq r < r_p$], and that the characteristic radius of the plasma column is much larger than the thermal Debye length, i.e., $r_p \gg \lambda_D \equiv (k_B T_e/4\pi\hat{n}_e e^2)^{1/2}$. Typical numerical results obtained from Eqs. (2) and (3) are illustrated in Fig. 1, where $n_{eq}(r)/\hat{n}_e$ is plotted versus r/λ_D for values of ϵ ranging from 10^{-2} to 10^{-6}. Note from Fig. 1 that the density drops abruptly to exponentially small values over a layer several Debye lengths in radial thickness. Furthermore, solving Eq. (4) for the angular rotation velocity ω_r gives two possible solutions, $\omega_r = \omega_r^+$ and $\omega_r = \omega_r^-$, where $\omega_r^\pm \equiv (\omega_c/2)\{1 \pm [1 - (2\hat{\omega}_p^2/\omega_c^2)(1+\epsilon)]^{1/2}\}$. Here, $\omega_r = \omega_r^+ > \omega_c/2$ corresponds to a "fast" rotational equilibrium, whereas $\omega_r = \omega_r^- < \omega_c/2$ corresponds to a "slow" rotational equilibrium. In Fig. 2, the solutions for ω_r/ω_c are plotted versus the dimensionless self-field parameter $2\hat{\omega}_p^2/\omega_c^2$ for the two cases $\epsilon = 0$ and $\epsilon = 0.2$. Note from Fig. 2 that the maximum allowed density consistent with a radially confined equilibrium is $[2\hat{\omega}_p^2/\omega_c^2]_{max} = (1+\epsilon)^{-1}$, in which case $\omega_r = \omega_r^\pm = \omega_c/2$.

For future reference, it is convenient to take the derivative of the expression for $n_{eq}(r)$ in Eq. (2) with respect to the radial coordinate r. Rearranging

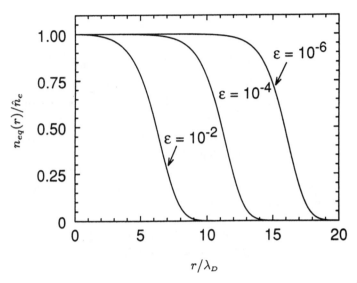

Fig. 1 Plots of normalized equilibrium density $n_{eq}(r)/\hat{n}_e$ versus r/λ_D calculated numerically from Eqs. (2) and (3). Each curve is specified by the
dimensionless parameter $\epsilon = (\omega_r \omega_c - \omega_r^2)/(\hat{\omega}_p^2/2) - 1$ [Eq. (4)] for values
of ϵ corresponding to $\epsilon = 10^{-2}$, 10^{-4}, and 10^{-6}.

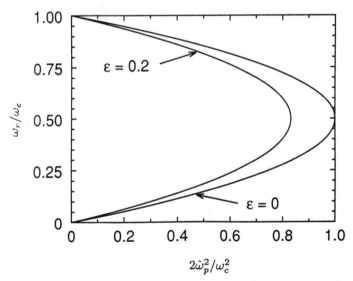

Fig. 2 Plots of normalized rotation frequency ω_r/ω_c versus the dimensionless
self-field parameter $2\hat{\omega}_p^2/\omega_c^2$ determined from Eq. (4) for the two cases
$\epsilon = 0$ and $\epsilon = 0.2$. The upper(lower) branch with $\omega_r/\omega_c > 1/2(\omega_r/\omega_c < 1/2)$ corresponds to a fast(slow) rotational equilibrium.

terms, this readily gives

$$-n_{eq}(r)m\omega_r^2 r + k_B T_e \frac{\partial}{\partial r} n_{eq}(r) = -en_{eq}(r)\left[E_r(r) + \frac{1}{c}\omega_r r B_0\right],\tag{5}$$

where $E_r(r) = -\partial\phi(r)/\partial r$ is the equilibrium radial electric field, and $E_r(r)$ solves the Poisson equation $r^{-1}(\partial/\partial r)[rE_r] = -4\pi en_{eq}$. As expected, Eq. (5) is simply a statement of equilibrium radial force balance on an electron fluid element in circumstances where the azimuthal flow velocity is $V_\theta(r) = \omega_r r$, and the scalar pressure is $P(r) = n_{eq}(r)k_B T_e$, where $T_e = const.$ Equation (5) will be used later in this article to calculate a closed analytical expression for the equilibrium mean-square radius $\langle r^2\rangle_{eq}$, as well as a power-series solution for the radial density profile $n_{eq}(r)$.

III. NONLINEAR CONSERVATION CONSTRAINTS

Several authors have examined the implications of the global (spatially-averaged)[1,19,20] and microscopic[1,26] conservation constraints for the cylindrical nonneutral plasma configuration considered here. For present purposes, we consider circumstances where the initial mean-square radius, $\langle r^2\rangle_{t=0} = [\int d^2x \int d^3p \, r^2 f(\mathbf{x},\mathbf{p},t=0)]/[\int d^2x \int d^3p \, f(\mathbf{x},\mathbf{p},t=0)]$, is small in comparison with the conducting wall radius-squared, i.e., $\langle r^2\rangle_{t=0} \ll r_w^2$. In this case, the radial loss of electrons to the wall is negligibly small during the nonlinear evolution of the system, and

$$N_e = \int d^2x \int d^3p \, f_e(\mathbf{x},\mathbf{p},t) = const.\tag{6}$$

and

$$\langle P_\theta\rangle = \frac{\int d^2x \int d^3p \, P_\theta f_e(\mathbf{x},\mathbf{p},t)}{\int d^2x \int d^3p \, f_e(\mathbf{x},\mathbf{p},t)} = const.\tag{7}$$

are exactly conserved quantities, where $P_\theta = r(p_\theta - mr\omega_c/2)$ is the canonical angular momentum. Here, $\int d^2x \cdots = \int_0^{2\pi} d\theta \int_0^{r_w} dr \, r \cdots$ is the perpendicular volume integral, $N_e = \int d^2x \int d^3p \, f_e = const.$ is the number of electrons per unit axial length, and $\langle P_\theta\rangle = N_e^{-1}(\int d^2x \int d^3p \, P_\theta f_e) = const.$ is the average canonical angular momentum per unit axial length.

Because N_e and $\langle P_\theta\rangle$ are exactly conserved quantities, we conclude that the $t = 0$ values of N_e and $\langle P_\theta\rangle$ are the same as the values of N_e and $\langle P_\theta\rangle$ in thermal equilibrium. Therefore, from Eqs. (1) and (2) we obtain

$$N_e = 2\pi\hat{n}_e \int_0^{r_w} dr \, r \, \exp\left\{-\frac{m}{2k_B T_e}\left[r^2(\omega_r\omega_c - \omega_r^2) - \frac{2e}{m}\phi(r)\right]\right\},\tag{8}$$

and

$$\langle P_\theta \rangle = m(\omega_r - \omega_c/2)\langle r^2 \rangle_{eq} . \tag{9}$$

Here, the mean-square radius in thermal equilibrium is defined by

$$\langle r^2 \rangle_{eq} = \frac{\int_0^{r_w} dr\, r\, r^2 \exp\{-(m/2k_BT_e)[r^2(\omega_r\omega_c - \omega_r^2) - (2e/m)\phi(r)]\}}{\int_0^{r_w} dr\, r\, \exp\{-(m/2k_BT_e)[r^2(\omega_r\omega_c - \omega_r^2) - (2e/m)\phi(r)]\}} , \tag{10}$$

and we take $r_w \to \infty$ in Eqs. (8) and (10) because $\langle r^2 \rangle_{t=0} \ll r_w^2$ is assumed. The global conservation constraints in Eqs. (6) and (7) [which lead to Eqs. (8) and (9)] are adequate for our purposes here, and use will not be made in the present analysis of the conservation of total energy, i.e., $\int d^2x \, \{|\nabla\phi(\mathbf{x},t)|^2/8\pi + \int d^3p\,(\mathbf{p}^2/2m)f_e(\mathbf{x},\mathbf{p},t) \} = const.$ Because $\langle r^2 \rangle_{eq} > 0$ in Eq. (9), it is important to note that $\omega_r < \omega_c/2$ whenever $\langle P_\theta \rangle < 0$ (which corresponds to a "slow" rotational equilibrium), whereas $\omega_r > \omega_c/2$ whenever $\langle P_\theta \rangle > 0$ (which corresponds to a "fast" rotational equilibrium).

IV. MEAN-SQUARE RADIUS IN THERMAL EQUILIBRIUM

In Eq. (5), the radial electric field $E_r(r) = -\partial\phi(r)/\partial r$, calculated from the equilibrium Poisson equation $r^{-1}(\partial/\partial r)[rE_r] = -4\pi e n_{eq}$, is given by

$$E_r(r) = -\frac{4\pi e}{r} \int_0^r dr'\, r'n_{eq}(r') , \tag{11}$$

where $n_{eq}(r)$ is the thermal equilibrium density profile defined in Eq. (2). Substituting Eq. (11) into Eq. (5), the equilibrium force balance equation can be expressed as

$$\frac{k_BT_e}{m}\frac{\partial}{\partial r}n_{eq}(r) = -(\omega_r\omega_c - \omega_r^2)rn_{eq}(r) + \frac{4\pi e^2}{m}\frac{n_{eq}(r)}{r}\int_0^r dr'\, r'n_{eq}(r') . \tag{12}$$

Equation (12) is a nonlinear integro-differential equation for the equilibrium density profile $n_{eq}(r)$. Operating on Eq. (12) with $\int_0^{r_w} dr\, r^2 \cdots$, where $r_w \to \infty$, some algebraic manipulation gives a *closed* expression for the mean-square radius $\langle r^2 \rangle_{eq}$ in thermal equilibrium. We obtain

$$\langle r^2 \rangle_{eq} = \frac{1}{m(\omega_r\omega_c - \omega_r^2)}[2k_BT_e + e^2N_e] , \tag{13}$$

where $N_e = 2\pi \int_0^\infty dr\, rn_{eq}(r)$ is the number of electrons per unit axial length, and $r_w \to \infty$ has been assumed. Note that Eq. (13) expresses $\langle r^2 \rangle_{eq}$ directly in terms of N_e, the angular rotation velocity ω_r, and the electron temperature T_e.

It is useful to keep in mind that in the cold-fluid limit with ω_r and T_e satisfying [see Eq. (4)]

$$\omega_r\omega_c - \omega_r^2 \;\rightarrow\; \frac{4\pi\hat{n}_e e^2/m}{2} = \frac{\hat{\omega}_p^2}{2}\,,$$

$$T_e \;\rightarrow\; 0\,, \tag{14}$$

the thermal equilibrium density profile $n_{eq}(r)$ in Eq. (2) becomes *rectangular*, with $n_{eq}(r) = \hat{n}_e$ for $0 \le r < r_0$, and $n_{eq}(r) = 0$ for $r > r_0$. Here, r_0 is the $T_e \to 0$ plasma radius defined by

$$\hat{n}_e\pi r_0^2 \equiv N_e\,. \tag{15}$$

This is illustrated in Fig. 3, where the normalized density $n_{eq}(r)/\hat{n}_e$ calculated numerically from Eqs. (2) and (3) is plotted versus r/r_0 for fixed N_e and $r_0 = (N_e/\hat{n}_e\pi)^{1/2}$, and values of λ_D^2/r_0^2 ranging from 0 to 0.1, corresponding to increasing values of electron temperature T_e.

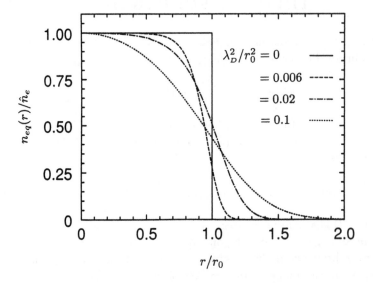

Fig. 3   Plots of normalized equilibrium density $n_{eq}(r)/\hat{n}_e$ versus r/r_0 calculated numerically from Eqs. (2) and (3) for fixed N_e and $r_0 = (N_e/\hat{n}_e\pi)^{1/2}$, and values of λ_D^2/r_0^2 ranging from 0 to 0.01.

Returning to Eq. (13), the full expression for $\langle r^2 \rangle_{eq}$ can be expressed in the equivalent form

$$\langle r^2 \rangle_{eq} = \frac{\hat{\omega}_p^2/2}{\omega_r \omega_c - \omega_r^2} \left[4\lambda_D^2 + \frac{1}{2} r_0^2 \right] . \tag{16}$$

In Eq. (16), $\lambda_D^2 \equiv k_B T_e / 4\pi \hat{n}_e e^2$ is the thermal Debye length-squared, and $r_0 \equiv (N_e/\hat{n}_e\pi)^{1/2}$ is defined in Eq. (15). In the special limiting case of a cold plasma with $T_e \to 0$ ($\lambda_D^2 \to 0$) and $\omega_r \omega_c - \omega_r^2 \to \hat{\omega}_p^2/2$, the expression for $\langle r^2 \rangle_{eq}$ in Eq. (16) reduces to $\langle r^2 \rangle_{eq} = r_0^2/2$, which is the expected result for a uniform density plasma column with density $\hat{n}_e$ and radius r_0, where $\langle r^2 \rangle_{eq} = (\int_0^{r_0} dr\, r\, r^2 \hat{n}_e)/(\int_0^{r_0} dr\, r \hat{n}_e) = r_0^2/2$. It should be noted that Eqs. (13) and (16) are equivalent representations of the mean-square radius $\langle r^2 \rangle_{eq}$, valid for general values of T_e and ω_r consistent with the condition $\omega_r \omega_{ce} - \omega_r^2 \geq \hat{\omega}_p^2/2$, required for radial confinement of the plasma column.

For fixed N_e and $r_0 = (N_e/\pi \hat{n}_e)^{1/2}$, it should be emphasized that λ_D^2/r_0^2 and the dimensionless parameter $\epsilon = (\omega_r \omega_c - \omega_r^2)/(\hat{\omega}_p^2/2) - 1$ defined in Eq. (4) are implicitly connected through the solution to Poisson's equation (3). This is illustrated in Fig. 4, which shows a *universal* plot of λ_D^2/r_0^2 versus ϵ, obtained by numerical solution to Poisson's equation (3) at fixed N_e and r_0. Consistent with Eq. (16) and the connection between λ_D^2/r_0^2 and ϵ illustrated in Fig. 4, shown in Fig. 5 is a plot of the normalized mean-square radius $\langle r^2 \rangle_{eq}/r_0^2$ versus λ_D^2/r_0^2, which is a dimensionless measure of the plasma temperature T_e. Note from Fig. 5 that $\langle r^2 \rangle_{eq} \to r_0^2/2$ in the cold-plasma limit where $\lambda_D^2 \to 0$, as expected.

For future reference, we combine Eqs. (9) and (13) and find that the average canonical angular momentum can also be expressed as

$$\langle P_\theta \rangle = \frac{\omega_r - \omega_c/2}{\omega_r \omega_c - \omega_r^2} [2 k_B T_e + e^2 N_e] , \tag{17}$$

which relates $\langle P_\theta \rangle$ directly to ω_r, T_e, and N_e.

V. ALTERNATE REPRESENTATIONS OF THE ANGULAR ROTATION VELOCITY AND THE MEAN-SQUARE RADIUS

Equations (13) and (17) constitute useful analytical expressions because they directly relate the quantities ω_r, $\langle r^2 \rangle_{eq}$, and T_e, which are properties of the thermal equilibrium plasma, to N_e and $\langle P_\theta \rangle$, which are globally conserved quantities during the nonlinear evolution of the system. In this regard, we

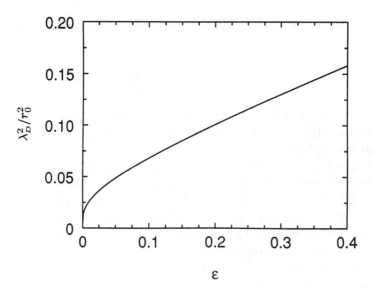

Fig. 4 Universal plot of $\lambda_D^2/r_0^2 = \pi\lambda_D^2\hat{n}_e/N_e$ versus $\epsilon = (\omega_r\omega_c - \omega_r^2)/(\hat{\omega}_p^2/2) - 1$ determined numerically from Eqs. (2) and (3).

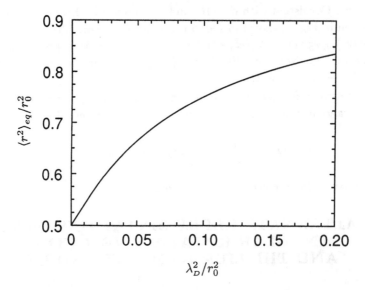

Fig. 5 Universal plot of $\langle r^2\rangle_{eq}/r_0^2$ versus $\lambda_D^2/r_0^2 = \pi\lambda_D^2\hat{n}_e/N_e$ determined from Eq. (16) and Fig. 4.

may wish to view N_e and $\langle P_\theta \rangle$ as known quantities [specified by the initial condition $f_e(\mathbf{x}, \mathbf{p}, t = 0)$, say], and T_e as a known quantity (a "measured" property of the thermal equilibrium plasma, say). Then Eqs. (13) and (17) can be used to obtain closed analytical expressions for the angular rotation velocity ω_r and the mean-square radius $\langle r^2 \rangle_{eq}$ directly in terms of N_e, $\langle P_\theta \rangle$, and T_e.

It is useful to introduce the dimensionless parameter $b(N_e, \langle P_\theta \rangle, T_e)$ defined by

$$
\begin{aligned}
b &= \frac{1}{\omega_c \langle P_\theta \rangle} [2k_B T_e + e^2 N_e] \\
&= \frac{m\hat{\omega}_p^2}{2\omega_c \langle P_\theta \rangle} \left[4\lambda_D^2 + \frac{1}{2} r_0^2 \right] ,
\end{aligned}
\tag{18}
$$

where $\hat{\omega}_p^2 \equiv 4\pi \hat{n}_e e^2 / m$, $\pi \hat{n}_e r_0^2 \equiv N_e$, and $\lambda_D^2 \equiv k_B T_e / 4\pi \hat{n}_e e^2$. Solving Eq. (17) for the angular velocity ω_r gives

$$
\omega_r = \omega_r^\pm \equiv \frac{\omega_c}{2} \left\{ 1 \pm \left[(1 + b^2)^{1/2} - |b| \right] \right\} ,
\tag{19}
$$

where the upper (lower) sign corresponds to an equilibrium with positive (negative) value of $\langle P_\theta \rangle$, i.e., $\langle P_\theta \rangle > 0$ or $\langle P_\theta \rangle < 0$, respectively. Figure 6 shows plots of the normalized frequencies ω_r^+/ω_c and ω_r^-/ω_c versus $|b|$ calculated from Eq. (19). Note that the information in Fig. 6 is fully equivalent to that in Fig. 2. Substituting Eq. (19) into Eq. (13), we obtain for the mean-square radius $\langle r^2 \rangle_{eq}$

$$
\langle r^2 \rangle_{eq} = \frac{2|\langle P_\theta \rangle|}{m\omega_c [(1 + b^2)^{1/2} - |b|]} .
\tag{20}
$$

Note also from Eq. (18) that Eq. (20) can be expressed in the equivalent form

$$
\begin{aligned}
\langle r^2 \rangle_{eq} &= \frac{2}{m\omega_c^2 |b| [(1 + b^2)^{1/2} - |b|]} [2k_B T_e + e^2 N_e] \\
&= \frac{\hat{\omega}_p^2}{\omega_c^2 |b| [(1 + b^2)^{1/2} - |b|]} \left[4\lambda_D^2 + \frac{1}{2} r_0^2 \right] .
\end{aligned}
\tag{21}
$$

Equations (19) and (20) constitute closed expressions for the thermal equilibrium quantities ω_r and $\langle r^2 \rangle_{eq}$ directly in terms of N_e, $\langle P_\theta \rangle$, and T_e.

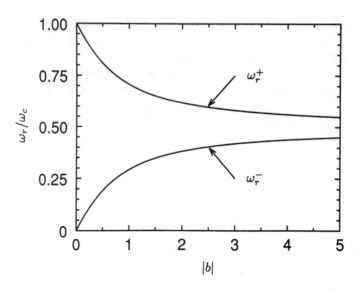

Fig. 6 Plots of the normalized rotation frequencies ω_r^+/ω_c and ω_r^-/ω_c versus $|b|$ calculated in Eq. (19). The dimensionless parameter b is defined by Eq. (18), and the ω_r^+ (ω_r^-) branch corresponds to $b > 0$ ($b < 0$), or equivalently, $\langle P_\theta \rangle > 0$ ($\langle P_\theta \rangle < 0$).

VI. POWER-SERIES SOLUTION FOR THE EQUILIBRIUM DENSITY PROFILE

We now return to the radial force balance equation (12) and introduce the scaled density variable $N(r) \equiv n_{eq}(r)/\hat{n}_e$ and the positive dimensionless quantity ϵ defined in Eq. (4) by $1 + \epsilon = (\omega_r \omega_c - \omega_r^2)/(\hat{\omega}_p^2/2)$. Equation (12) then becomes

$$\lambda_D^2 \frac{\partial}{\partial r} N(r) = -\frac{1}{2}(1 + \epsilon)rN(r) + \frac{N(r)}{r} \int_0^r dr' \, r'N(r') \, , \qquad (22)$$

where $\lambda_D^2 = k_B T_e/4\pi\hat{n}_e e^2$. Examination of Eq. (22) shows that $N(r)$ is naturally scaled by $1 + \epsilon$ and r^2 by $\lambda_D^2/(1 + \epsilon)$. Therefore, we define

$$X = \frac{r^2}{r_0^2} \, , \qquad r_0^2 = \frac{4\lambda_D^2}{1 + \epsilon} \, , \qquad Y(X) = \frac{N(x)}{1 + \epsilon} \, , \qquad (23)$$

and Eq. (22) can be expressed in the equivalent form

$$X \frac{\partial}{\partial X} Y(X) = -XY(X) + Y(X) \int_0^X dX' \, Y(X') \, . \qquad (24)$$

The solution to Eq. (24) can be expressed in the form of a power series with

$$Y(X) = \sum_{n=0}^{\infty} \alpha_n X^n , \tag{25}$$

where $Y(X = 0) = N(X = 0)/(1+\epsilon) = (1+\epsilon)^{-1}$. Substituting Eq. (25) into Eq. (24) and equating the coefficients of like powers of X readily gives

$$\alpha_0 = (1+\epsilon)^{-1} ,$$
$$\alpha_1 = -\alpha_0 + \alpha_0^2 ,$$
$$\alpha_2 = -\frac{1}{2}\alpha_1 + \frac{3}{4}\left(\frac{1}{2}\alpha_0\alpha_1 + \frac{1}{2}\alpha_1\alpha_0\right) = \frac{1}{2}(\alpha_0 - \alpha_0^2) - \frac{3}{4}\alpha_0(\alpha_0 - \alpha_0^2) \tag{26}$$
$$\vdots$$
$$\alpha_{n+1} = -\frac{\alpha_n}{n+1} + \frac{n+2}{2(n+1)} \sum_{m=0}^{n} \frac{\alpha_m \alpha_{n-m}}{(m+1)(n-m+1)} .$$

From Eqs. (23) and (25), the equilibrium electron density $n_{eq} = \hat{n}_e N = \hat{n}_e(1+\epsilon)Y$ is given by

$$n_{eq}(r) = \hat{n}_e(1+\epsilon) \sum_{n=0}^{\infty} \alpha_n \left[\frac{(1+\epsilon)r^2}{4\lambda_D^2}\right]^n . \tag{27}$$

Note that the recursion formula (26) can be used to express α_n, $n \geq 2$, directly in terms of $\alpha_1 = -\alpha_0 + \alpha_0^2 = -\epsilon/(1+\epsilon)^2$. Moreover, as expected, for *small* values of $\epsilon \ll 1$, the plasma column is radially broad in comparison with the thermal Debye length λ_D, and the plasma density $n_{eq}(r)$ is approximately constant (and equal to $\hat{n}_e$) in the column interior.

The closed analytical expression for $n_{eq}(r)$ in Eq. (27) is useful for testing the numerical solution to Eqs. (2) and (3), at least in the small-r region near the axis of the plasma column. For example, while $[\partial n_{eq}/\partial r]_{r=0} = 0$ follows from Eq. (27), it is readily shown that $[\partial^2 n_{eq}/\partial r^2]_{r=0} = -\epsilon\hat{n}_e/2\lambda_D^2$, corresponding to a weak downward concavity in the density profile whenever $\epsilon \ll 1$.

VII. CONCLUSIONS

In this paper, we have calculated the thermal equilibrium properties of a cylindrical, pure electron plasma confined radially by a uniform axial magnetic field $B_0 \mathbf{e}_z$. Following a discussion of the equilibrium model in the weak coupling approximation (Sec. II), the nonlinear conservation constraints, $N_e = const.$ [Eq. (6)] and $\langle P_\theta \rangle = const.$ [Eq. (7)], were used to derive closed analytical expressions for the equilibrium mean-square radius $\langle r^2 \rangle_{eq}$ of the plasma column and the equilibrium angular rotation velocity ω_r, expressed in terms of N_e, $\langle P_\theta \rangle$, and the electron temperature T_e (Secs. III–V). The detailed shape of the equilibrium density profile $n_{eq}(r) = \int d^3p\, f_{eq}$ was characterized over a wide range of system parameters (Figs. 1 and 3), and universal plots were obtained which relate the normalized temperature λ_D^2/r_0^2 to the dimensionless parameter $\epsilon = (\omega_r \omega_c - \omega_r^2)/(\hat{\omega}_p^2/2) - 1$ (Fig. 4), and the normalized mean-square radius $\langle r^2 \rangle_{eq}/r_0^2$ to λ_D^2/r_0^2 (Fig. 5). Finally, in Sec. VI, an exact power-series solution was derived for the thermal equilibrium density profile $n_{eq}(r)$ as a function of the radial distance r from the axis of the plasma column [Eq. (27)].

ACKNOWLEDGMENTS

This research was supported by the Office of Naval Research.

REFERENCES

[1] R. C. Davidson, *Physics of Nonneutral Plasmas* (Addison-Wesley, Reading, MA, 1990), and references therein.

[2] *Non-neutral Plasma Physics*, edited by C. W. Roberson and C. F. Driscoll (American Institute of Physics, New York, 1988), and papers therein.

[3] C. F. Driscoll and K. S. Fine, Phys. Fluids B **2**, 1359 (1990).

[4] R. A. Smith and T. M. O'Neil, Phys. Fluids B **2**, 2961 (1990).

[5] R. C. Davidson, H.-W. Chan, C. Chen, and S. M. Lund, Rev. Mod. Phys. **63**, 341 (1991).

[6] T. M. O'Neil and R. A. Smith, Phys. Fluids B **4**, 2721 (1992).

[7] S. M. Lund, J. J. Ramos, and R. C. Davidson, Phys. Fluids B **5**, in press (1993).

[8] C. Chen and R. C. Davidson, Phys. Rev. A **42**, 5041 (1990).

[9] C. Chen and R. C. Davidson, Phys. Rev. A **43**, 5541 (1991).

[10] J. H. Malmberg and T. M. O'Neil, Phys. Rev. Lett. **39**, 1333 (1977).

[11] G. C. Grimes and G. Adams, Phys. Rev. Lett. **42**, 795 (1979).

[12]D. H. E. Dubin and T. M. O'Neil, Phys. Rev. Lett. **60**, 511 (1988).

[13]S. L. Gilbert, J. J. Bollinger and D. J. Wineland, Phys. Rev. Lett. **60**, 2022 (1988).

[14]J. S. deGrassie and J. H. Malmberg, Phys. Fluids **23**, 63 (1980).

[15]G. Rosenthal, G. Dimonte, and A. Y. Wong, Phys. Fluids **30**, 3257 (1987).

[16]T. M. O'Neil, Phys. Fluids **26**, 2128 (1983).

[17]T. M. O'Neil and C. F. Driscoll, Phys. Fluids **22**, 266 (1979).

[18]C. F. Driscoll, J. H. Malmberg, and K. S. Fine, Phys. Rev. Lett. **60**, 1290 (1988).

[19]R. C. Davidson and N. A. Krall, Phys. Rev. Lett. **22**, 833 (1969).

[20]R. C. Davidson and N. A. Krall, Phys. Fluids **13**, 1543 (1970).

[21]C. Michael, Proc. Astronomical Society of Australia **6**, 127 (1985).

[22]G. Gabrielse, X. Fei, K. Helmerson, S. L. Rolston, R. Tjoelker, T. A. Trainor, H. Kalinowsky, J. Haas, and W. Kells, Phys. Rev. Lett. **57**, 2504 (1986).

[23]G. Gabrielse, X. Fei, L. A. Orozco, R. L. Tjoelker, J. Haas, H. Kalinowsky, T. A. Trainor, and W. Kells, Phys. Rev. Lett. **65**, 1317 (1990).

[24]C. M. Surko and T. J. Murphy, Phys. Fluids B **2**, 1372 (1990).

[25]R. C. Davidson, Phys. Fluids **27**, 1804 (1984).

[26]T. M. O'Neil, Phys. Fluids **23**, 2216 (1980).

[27]T. H. Stix, *Theory of Plasma Waves* (American Institute of Physics, New York, 1992).

[28]S. A. Prasad and T. M. O'Neil, Phys. Fluids **22**, 278 (1979).

INTRODUCTION TO ULTRA-INTENSE LASER-PLASMA INTERACTIONS*

W. L. Kruer and S. C. Wilks
Lawrence Livermore National Laboratory, Livermore, CA 94550

ABSTRACT

A brief introduction to the interaction of ultra-intense laser pulses with plasmas is presented. A number of interesting phenomena due to relativistic electron dynamics are reviewed in simple calculations. These relativistic effects range from nonlinear frequency shifts of light waves to penetration of overdense plasmas, filamentation and self-focusing. Finally, computer simulations are used to illustrate several strongly nonlinear effects in such plasmas, including heating by the oscillating ponderomotive force and relativistic filamentation.

* Work performed under the auspices of the United States Department of Energy by the Lawrence Livermore National Laboratory under contract number W-7405-ENG-48.

INTRODUCTION

Table top terrawatt lasers[1] are providing access to a novel regime of laser-plasma interactions. Such lasers produce short pulses (pulse lengths of order 100 fs to 1 ps) with very high intensities. The intensity has now exceeded 10^{18} W/cm^2, with factors of 10-10^3 increase projected in the near future. The oscillatory velocity of electrons in such an intense pulse becomes nearly the velocity of light, introducing many effects associated with relativistic dynamics. Furthermore, the light pressure is extremely large; i.e., over 300 megabars when the intensity is 10^{18} W/cm^2. The interaction of these short, intense pulses with plasmas is now being addressed in a growing number of experiments[2-3] and calculations[4-5].

Here we give a brief introduction to this rich regime. Calculations are used to review some of the effects associated with relativistic electron motion. These effects range from nonlinear frequency shifts to relativistic filamentation and penetration of overdense plasmas. Finally, we conclude with computer simulations emphasizing strongly nonlinear effects in such plasmas.

PONDEROMOTIVE FORCE WITH RELATIVISTIC DYNAMICS

Let's begin by deriving a relativistically correct expression for the ponderomotive force.[6] The force equation for a warm electron fluid with velocity **u** and momentum **p** is

$$\frac{\partial \mathbf{p}}{\partial t} + \mathbf{u} \cdot \nabla \mathbf{p} = -e \left[\mathbf{E} + \frac{\mathbf{u} \times \mathbf{B}}{c} \right], \tag{1}$$

where **E** is the electric field and **B** the magnetic field. In the usual way, we introduce the vector potential **A** and electrostatic potential ϕ:

$$\mathbf{B} = \nabla \times \mathbf{A},$$
$$\mathbf{E} = -\frac{1}{c} \frac{\partial \mathbf{A}}{\partial t} - \nabla \phi. \tag{2}$$

Since $\mathbf{p} = \gamma m_0 \mathbf{u}$, we can express $\gamma = \sqrt{1 + \frac{p^2}{m_0^2 c^2}}$, where m_0 is the rest mass of an electron. Eq. 1 becomes

$$\frac{\partial \mathbf{p}}{\partial t} + \frac{\mathbf{p} \cdot \nabla \mathbf{p}}{m_0 c \sqrt{1 + \frac{p^2}{m_0^2 c^2}}} = -e \left[-\frac{1}{c} \frac{\partial \mathbf{A}}{\partial t} - \nabla \phi + \frac{\mathbf{p}}{m_0 c \sqrt{1 + p^2/m_0^2 c^2}} \times \nabla \times \mathbf{A} \right] \tag{3}$$

We next separate the momentum into transverse ($\mathbf{p_t}$) and longitudinal ($\mathbf{p}_\ell$) components; i.e. $\mathbf{p} = \mathbf{p_t} + \mathbf{p}_\ell$. The transverse component of Eq. 3 yields $\mathbf{p_t} = e\mathbf{A}/c$. This oscillatory momentum in the transverse field normalized to $m_0 c$ is a very useful parameter, which can be written as

$$\frac{|\mathbf{p_t}|}{m_o c} = \left(\frac{I\lambda_\mu^2}{1.3\times10^{18}}\right)^{1/2} .$$

Here I is the free-space intensity of the light wave in W/cm^2 and λ_μ is its wavelength in microns. Note that when $I\lambda_\mu^2 \geq 10^{18} \frac{\text{W-}\mu\text{m}^2}{\text{cm}^2}$, the oscillatory velocity becomes nearly c.

The longitudinal component of Eq. 3. gives

$$\frac{\partial \mathbf{p}_\ell}{\partial t} = e\nabla\phi - m_o c^2 \nabla\sqrt{1+\frac{p^2}{m_o^2 c^2}} , \tag{4}$$

where we have noted that $e \nabla \times \mathbf{A} = c\nabla \times \mathbf{p}$, since $\nabla \times \mathbf{p}_\ell = 0$. The following vector identity was also used:

$$\frac{\mathbf{a}\cdot\nabla\mathbf{a}}{\sqrt{1+\mathbf{a}^2}} + \frac{\mathbf{a}\times\nabla\times\mathbf{a}}{\sqrt{1+\mathbf{a}^2}} = \nabla\sqrt{1+\mathbf{a}^2} ,$$

where $\mathbf{a}$ is any vector. The second term on the right hand side of Eq. 4 is the ponderomotive force ($\mathbf{f_p}$), which can also be expressed as

$$\mathbf{f_p} = -\nabla \left\{(\gamma-1) m_o c^2\right\} , \tag{5}$$

where $\gamma = \sqrt{1+\frac{p^2}{m_o^2 c^2}}$. Note that the ponderomotive force is simply the gradient of the oscillatory energy of an electron in the transverse and longitudinal fields. In the non-relativistic limit, Eq. 5 reduces to the well-known expression $\mathbf{f_p} = -\nabla \frac{m_o v_{os}^2}{2}$, where v_{os} is the electron oscillatory velocity.

LIGHT WAVE PROPAGATION

Let us now consider the propagation of light waves in a uniform plasma with density n_o including relativistic electron dynamics. For simplicity, we take a circularly polarized wave:

$$\mathbf{A} = A_o\left[\cos\left(kz-\omega t\right)\hat{x} + \sin\left(kz-\omega t\right)\hat{y}\right] \tag{6}$$

As can be seen from Eq. 4, such a wave is purely transverse, i.e. $\mathbf{p}_\ell = 0$ and both $|\mathbf{p_t}|^2$ and γ are independent of time.

Substituting $\mathbf{B} = \nabla \times \mathbf{A}$ into Faraday's law gives

$$\left(\frac{\partial^2}{\partial t^2} - c^2\nabla^2\right)\mathbf{A} = 4\pi c\, \mathbf{J_t}\ , \tag{7}$$

where $\mathbf{J_t}$ is the transverse current density:

$$\mathbf{J_t} = -n_o e\, \mathbf{u_t} = \frac{-n_o e\, \mathbf{p_t}}{m_o\sqrt{1+\dfrac{\mathbf{p_t^2}}{m_o^2 c^2}}} \tag{8}$$

Since $\mathbf{p_t} = e\mathbf{A}/c$, we readily obtain the wave equation

$$\left(\frac{\partial^2}{\partial t^2} - c^2\nabla^2\right)\mathbf{A} = \frac{-\omega_{pe}^2 \mathbf{A}}{\sqrt{1+\left(\dfrac{e\mathbf{A}}{m_o c^2}\right)^2}}\ . \tag{9}$$

Here ω_{pe} is the electron plasma frequency: $\omega_{pe} = (4\pi n_o e^2/m_o)^{1/2}$.

Substitution of Eq. 6 into Eq. 9 gives the dispersion relation

$$\omega^2 = k^2 c^2 + \frac{\omega_{pe}^2}{\sqrt{1+\left(\dfrac{eA_o}{m_o c^2}\right)^2}} \tag{10}$$

The plasma frequency is effectively decreased due to the relativistic mass increase. One interesting consequence is apparent. A light wave can now propagate to a higher plasma density. The so-called critical density is increased[7] by the factor $\sqrt{1+\left(\dfrac{eA_o}{m_o c^2}\right)^2}$.

Linearly-polarized waves are not purely transverse. As shown by Eq. 4, there is now a longitudinal component, which drives a density fluctuation. The nonlinear dynamics is much richer. It can be shown that the dispersion relation now becomes[8]

$$\omega^2 = k^2 c^2 + \frac{\omega_{pe}^2}{\sqrt{1+\dfrac{1}{2}\left(\dfrac{eA_o}{m_o c^2}\right)^2}}\ , \tag{11}$$

where we assume that $\omega \gg \omega_{pe}$. Note that there is again a nonlinear frequency shift, since the electron mass depends on intensity.

RELATIVISTIC FILAMENTATION AND SELF-FOCUSING

The dependence of the frequency of a light wave on intensity introduces filamentation and self-focusing. The spatial growth rate for relativistic filamentation can be obtained by a straight-forward modification of the well-known results for (non-relativistic) ponderomotive filamentation. In the latter case[9] and in the quasi-static limit, the intensity-dependent change (δn_e) in electron density is

$$\frac{\delta n_e}{n_o} = -\frac{1}{4}\left(\frac{v_{os}}{v_e}\right)^2. \tag{12}$$

Here $v_{os} = eA_0/m_0c$ is the oscillatory velocity of an electron in the laser field, v_e the electron thermal velocity, and n_0 the background electron density. If we refer to Eq. 11, the effective density change due to the relativistic mass correction is

$$\frac{\delta n_e}{n_o} = -\frac{1}{4}\left(\frac{v_{os}}{c}\right)^2 \tag{13}$$

where we are assuming a linearly polarized wave with $(v_{os}/c)^2 \ll 1$.

Comparison of Eqs. 12 and 13 shows how to map the results for ponderomotive filamentation into predictions for relativistic filamentation. In particular, we simply replace v_{os}/v_e by v_{os}/c. The most unstable filament then has a transverse wave number k_m given by

$$k_m = \frac{1}{2}\frac{\omega_{pe}}{c}\frac{v_{os}}{c}. \tag{14}$$

The maximum spatial growth rate (k_i) is

$$k_i = \frac{1}{8}\left(\frac{v_{os}}{c}\right)^2\frac{\omega_{pe}^2}{k_oc^2}. \tag{15}$$

These results agree with early work.[10]

The same physical process leads to whole beam self-focusing. The spatial variation is now a zero-order one due to the finite width of the laser beam (which we assume to have a Gaussian intensity distribution with initial radius r_0). The nonlinear focusing is straight-forward to understand. Inside the beam, the effective plasma density is reduced by the relativistic mass correction. Since light waves are refracted towards low density, the beam width narrows and the intensity increases, leading to an even stronger focusing. Detailed calculations need to account for the fact that the intensity distribution does not remain Gaussian and that different parts of the beam focus at different places.[11] However, an effective focusing length ℓ_{SF} can be simply estimated as the growth length of a filament with $k \cong \frac{\pi}{2r_0}$. For relativistic self-focusing in the limit $\left(\frac{v_{os}}{c}\right)^2 \ll 1$,

$$\ell_{SF} \cong 2r_0\frac{\omega_0}{\omega_{pe}}\frac{c}{v_{os}}. \tag{16}$$

We've assumed that r_0 is much greater than the wavelength of the most unstable filament.

Self-focusing clearly requires that ℓ_{SF} be less the ℓ_{DIF}, the distance over which the beam diffracts. Since $\ell_{DIF} \cong \dfrac{\pi r_0^2}{\lambda_o}$, a threshold laser power (P_{th}) is obtained:[5]

$$P_{th} \cong 17\frac{n_{cr}}{n} \, GW , \tag{17}$$

where n_{cr} is the critical density defined by $\omega_{pe} = \omega_o$.

COMPUTER SIMULATIONS

These various calculations provide insight into some of the many rich effects introduced by relativistic plasma dynamics in ultra-high laser fields. Let's now consider computer simulations to illustrate more highly nonlinear and/or kinetic phenomena. We will use a 2-1/2 D particle code[12] which allows three velocity vectors but variations in only two directions. In the simulations linearly polarized laser light is propagated from vacuum onto a slab of plasma. The normally incident light wave can have its electric vector in or out of the simulation plane, i.e., the plane in which variations are followed.

Let's consider an example which illustrates phenomena with both underdense and overdense plasma. As shown in Fig. 1, the initial plasma density varies from .2 n_{cr} to .6 n_{cr} in a distance of about 10λ_o and then rises rapidly to 3n_{cr}. We take the ions as fixed (i.e., infinitely massive) in this simulation. Periodic boundary conditions are used in the transverse direction. Electrons striking the right boundary are re-emitted with their initial thermal temperature, mocking up contact with a higher density plasma. The incident light is a plane wave with $eA_o/mc^2 = .8$ and with its electric vector out of the simulation plane.

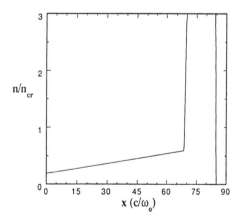

Figure 1: The initial plasma density profile

As the light wave propagates through the underdense plasma, it breaks into intense filaments. These filaments are apparent in Fig. 2, which is a contour plot of the transverse electric field at $\omega_0 t = 480$. This time would correspond to about 240 fs for a 1.06 μm light wave. The narrow regions of intense field have a full width of about 6-10 c/ω_0, and the peak field amplitude is $eA_0/mc^2 \cong 2.5$. Reflection of the light wave by the overdense plasma (which begins at $\omega_0 x/c \cong 70$) is also clearly visible.

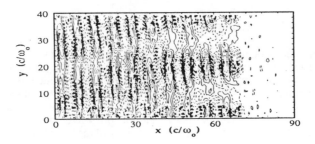

Figure 2: A contour plot of the amplitude of the transverse electric field at $\omega_0 t = 480$. The incident wave amplitude was $\dfrac{eA_0}{mc^2} = .8$

Note that the ponderomotive force is sufficient to generate strong charge separation in intense, narrow filaments. If we balance the ponderomotive potential ($e\phi_p$) with that due to charge separation, we estimate

$$\frac{\delta n_e}{n} \sim \frac{e\phi_p}{m_0 \omega_{pc}^2 \ell^2}.$$

Here δn_e is the change in electron density, n_0 the background density, and ℓ the half-width of the filaments. For $e\phi_p \gtrsim m_0 c^2$ and $\ell \sim 3\, \omega_{pe}/c$, $\delta n_e/n$ is of order one.

Fig. 3 shows an electron phase space (p_x versus x, where x is the direction of propagation of the light wave). There is significant electron heating in both the underdense plasma and at the interface with the overdense plasma. In this simulation, the heated electrons transiting into the overdense plasma have an effective temperature of about 330 keV. For these intense fields, so-called J x B heating[13-14] is significant at the overdense surface. This heating is due to the oscillating component of the ponderomotive force (see Eq. 5.) which efficiently accelerates electrons entering the interface with the proper phases.

Of course, on longer time scales it becomes essential to include effects due to ion motion. Then ions are ejected from the filaments, and holes are punched into the overdense surface. Indeed, a pronounced rippling of the reflection surface is apparent in simulations with higher intensity light even without ion inertia. Fig. 4 shows a contour plot of the laser field from a simulation with identical parameters to the one just discussed, except the initial light wave amplitude is $eA_0/m_0c^2 = 3$. Note the strong penetration of the overdense plasma correlated with regions of intense fields as expected from the relativistic mass variations. In turn, reflection from this rippled surface leads to a very complex pattern of structures in the underdense plasma.

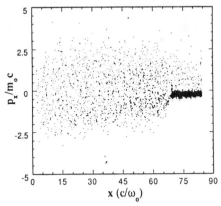

Figure 3: An electron phase space from the simulation at $\omega_0 t = 480$

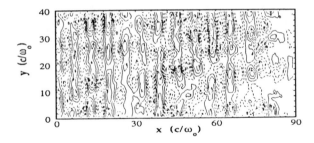

Figure 4:   A contour plot of the amplitude of the transverse electric field at $\omega_0 t = 480$. The incident wave amplitude was $\dfrac{eA_0}{mc^2} = 3$

Computer simulations[14] with ultra-intense pulses and finite ion inertia have identified many other important phenomena. Key features include generation of electrons and inward-directed ions with energies in the Mev range $\left(\text{for } I\lambda_\mu^2 \sim 10^{19} \text{ W–}\mu\text{m}^2/\text{cm}^2\right)$ and strong hole boring into the overdense plasma. The absorption is then enhanced by "not-so-resonant" absorption[4] on the sides of the holes. Self-generated magnetic fields of order 10^8 Gauss, a Rayleigh-Taylor like instability of the light-plasma interface, and significant Raman scattering are also found.

In summary, the interaction of ultra-intense laser pulses with plasmas is a challenging new area of research. Many important phenomena have been predicted, and many no doubt will be discovered as this regime is further addressed in experiments. Potential applications are numerous, including new radiation sources, particle accelerators and even a novel fusion scheme.[15]

ACKNOWLEDGMENTS:

We are grateful for valuable interactions with E. M. Campbell, C. Darrow, A. B. Langdon, W. Mori, M. Perry, M. Tabak and J. Woodworth. This work was performed under the auspices of the United States Department of Energy by the Lawrence Livermore National Laboratory under contract number W-7405-ENG-48.

REFERENCES

1. *Selected Papers on High Power Lasers,* edited by J. H. Soures (SPIE, Washington, 1991), p. 675-702,; G. Mourou and D. Umstadter, *Phys. Fluids* **B4**, 2315 (1992).

2. J. D. Kmetec et al., ibid **68**, 2309 (1992); C. S. Darrow et al., ibid **69**, 442 (1992).

3. X. Liu and D. Umstadter, *Phys. Rev. Lett.* **69**, 1935 (1992); M.M. Murnane, H. C. Kepleyn, M. D. Rosen and R. W. Falcone, *Science* **251**, 531 (1991).

4. F. Brunel, *Phys. Rev. Lett.* **59**, 52 (1987); P. Gibbon and H. R. Bell, ibid **68**, 1535 (1992); G. Bonnaud et al., *Laser and Particle Beams* **9**, 339 (1991).

5. G. Sun et al., *Phys. Fluids* **30**, 526 (1987); W. Mori et al., *Phys. Rev. Lett.* **60**, 1598 (1988); P. Sprangle et al., ibid **69**, 2200 (1992); T. Antonsen and P. Mora, ibid **69**, 2204 (1992).

6. C. J. McKinstrie and D. F. DuBois, *Phys. Fluids* **31**, 270 (1988).

7. P. K. Kaw and J. M. Dawson, *Phys. Fluids* **13**, 472 (1970).

8. A. I. Akhiezer and R. V. Polovin, *Sov. Phys.* JETP **3**, 696 (1956).

9. W. L. Kruer, *Comments Plasma Phys. Controlled Fusion* **9**, 63 (1985).

10. C. F. Max, J. Arons, and A. B. Langdon, *Phys. Rev. Lett.* **33**, 209 (1974); G. Schmidt and W. Horton, *Comments Plasma Phys. Controlled Fusion* **9**, 85 (1985).

11. B. I. Cohen et al., *Phys. Fluids* **B3**, 766 (1991).

12. A. B. Langdon and B. F. Lasinski, in *Methods in Computational Physics*, edited by J. Killeen, R. Alder, S. Fernbach, and M. Rotenberg (Academic, New York, 1976), Vol. 16, p. 327.

13. W. L. Kruer and K. G. Estabrook, *Phys. Fluids* **28**, 430 (1985).

14. S. C. Wilks, W. L. Kruer, M. Tabak and A. B. Langdon, *Phys. Rev. Lett.* **69**, 1383 (1992).

15. M. Tabak, et. al., private communication.

Two and Three Dimensional Analysis of Nonlinear Rayleigh-Taylor Instability

D. Ofer, J. Hecht, D. Shvarts
Physics Dept., Nuclear Research Centre Negev

Z. Zinamon
Nuclear Physics Dept., Weizmann Institute of Science

S.A. Orszag
Mechanical and Aerospace Engineering, Princeton University

R.L. McCrory
Laboratory for Laser Energetics, University of Rochester

Dedicated to Professor Thomas H. Stix on his 65th Birthday

1. Introduction

The linear stage of evolution of a single wavelength sinusoidal perturbation at a density-stratified interface is well known and understood. During the last decade, many studies have been devoted to the study of the late time evolution of the Rayleigh-Taylor (RT) instability. In the nonlinear stage, the perturbation evolves into a classical spike-bubble shape in which the bubble reaches a constant terminal velocity and the quickly spike achieves a nearly constant downward acceleration. The value of the Atwood ratio, $A = (\rho_1 - \rho_2)/(\rho_1 + \rho_2)$ where ρ_1 (ρ_2) is the density above (below) the interface, governs the basic physics. When the Atwood ratio is less than 1, there is an additional instability present that takes place along the sides of the spikes due to velocity shear. This instability, which is very similar to the classical Kelvin-Helmholtz instability, results in the late-time formation of "hammerhead" shaped spikes for which the constant acceleration phase gives way to constant velocity spikes at late times.

The aim of this paper is to report on some recent studies that were performed in order to further extend our understanding of the late time evolution of the RT instability. Here we report results on:

a)   Two-dimensional nonlinear interaction of a small number of modes;

b)   Three-dimensional single mode and two-mode-coupling evolution as compared with 2D nonlinear results;

c)   RT evolution in three-dimensional spherical geometry under conditions relevant to inertial confinement fusion (ICF) targets.

2. Code Description

Our 2D code is LEEOR2D is a compressible ALE (Arbitrary Lagrangian Eulerian) hydrodynamics code in which we have incorporated an Interface-Tracking (IT) scheme similar to that presented by Youngs [1]. The ALE scheme enables the code to take advantage of the benefits of both Lagrangian and Eulerian schemes. Each time-step is divided into two principal stages: (1) a Lagrangian stage in which mesh points flow with the physical material; and (2) a rezone Eulerian stage in which the Lagrangian mesh is remapped to a more desirable mesh (specified by the user). In order to allow for the presence of more than one material in a problem and large fluid distortions while still retaining the full flexibility' of the ALE method, we have added an IT step in the code. Each cell may contain an arbitrary number of materials. Basic material properties (such as density and internal energy) are stored separately, and the interface between materials in a 2D cell is assumed to be a straight line.

LEEOR3D is a direct generalization of LEEOR2D to three dimensions. Each computational cell is a 3D box with 8 vertices. The interface in the case of a mixed cell is a plane dividing the cell.

3. Mode-Coupling in Two Dimensions

We have previously studied [2] the interaction of a small number of modes in a two fluid RT instability at relatively late stages of development, i.e. the highly non-linear regime. We have identified strong interactions between modes that are both relatively long range in wave number space and also act in both directions, *viz.* short wavelengths affect long wavelengths and *visa versa*.

In studying the interaction of mode pairs, we identify three distinct stages in the development of the inter-fluid mixing region: (i) a linear stage, in which each mode develops independently of other modes; (ii) an "ordered" non-linear stage, in which strong, highly non-linear interactions between modes is present but the flow remains relatively ordered and specific modes can still be identified and quantified; and (iii) a "turbulent" stage, in which modal analysis (as defined in [2]) is no longer relevant and we chose instead to analyze statistical properties of the flow.

It has been found that during the second stage, mode pairs interact by suppressing the single mode growth rate, as compared with the growth rate of that mode when present alone. We have shown both how a long wavelength mode can suppress the growth of a short wavelength mode and *visa versa*. In each regime, the mechanism for suppression has been identified. We define the onset of suppression as the time when the suppressed mode has been reduced to 90% of the amplitude it would attain if it were growing alone with the same initial amplitude. We have found that: (i) the onset of suppression occurs at a constant amplitude of the suppressing mode, regardless of the suppressed mode amplitude; (ii) the suppressing amplitude (of the suppressing mode) is of the form

$$a_{suppression} = f(A)\sqrt{(\lambda_1 \lambda_2)}$$

where λ_1 (λ_2) is the wavelength of the suppressing (suppressed) mode, A is the Atwood ratio, and $f(A)$ changes smoothly from 0.05 when a finite-amplitude long wave suppresses the growth of a short wave to about 0.15 in the case of a short wave suppressing a long one (see Fig. 1).

For the case of a short wavelength suppressing a long wavelength mode, we have identified the cause of suppression to be a density gradient effect. The short-wavelength bubble-spike structure creates an effective density gradient in which the long wavelength must grow. This reduces the exponential growth rate γ of a wavenumber k mode according to $\gamma_{eff} = \sqrt{[Akg/(1 + kL)]}$, where L is the effective gradient scale length which is proportional to $a_{small}^2/\lambda_{small}$ (see [2]).

In the regime of long wavelength modes suppressing the growth of short waves, the cause of suppression, in our opinion, is a shear effect in which the long wavelength flow field causes the short wavelength structures to flow laterally instead of vertically, thereby reducing their growth (see Fig. 2). This effect has not been observed when A = 1 at least for the duration of our simulations. This behavior at A = 1 indicates that the main mechanism relevant to this phenomenon, at least for the time scales of our simulations, is probably the velocity shear, which is not present when A = 1, and not interface stretching. Comparison of the velocity of a linear-regime long wavelength mode to the non-linear-regime short wavelength velocity yields the empirical functional dependence of the onset for suppression on the wavelengths.

4. Multi-Mode Coupling

Of particular importance to ICF is the case where a broad band of short wavelength modes, representing the smaller scale laser and target inhomogeneities, appear with a discrete set of long wavelength modes, representing the large scale illumination nonuniformity. We have therefore extended our study of two mode coupling to the multi-mode case, especially cases in which a band of short wavelength modes interacts with a single long wavelength mode. In Fig. 3, we plot results for such a case in which the initial modes are $\ell = 2$ and a band of $\ell = 11\text{-}20$, at both intermediate and late stages of flow development. When more than one short wavelength mode is initially present, the analysis should distinguish between two distinct mode interaction mechanisms: (i) the generation of a long wavelength mode through the interaction of two short wavelengths, as discussed by Haan [3]; (ii) the nonlinear suppression of the long wavelength mode by the density gradient induced by the short wavelengths, as described above [2].

As regards to the first mechanism, we have found that Haan's second order mode coupling term (in which two modes generate additional modes with $\ell = \ell_1 + \ell_2$) can be extended much beyond his original estimates. We found that the formula suggested by Haan for weakly non-linear interactions can be used up to at least short wavelength amplitudes comparable to λ_{small} and not merely $\sim 0.1\ \lambda_{small}$. We believe that this formula is reasonably accurate so long as the long wavelength amplitude is small compared to its own wavelength.

After subtracting the contribution of the first mechanism to the long wavelength amplitude (including the correct phase), one is left with the effects of the second mechanism. We have found that one can generalize the above results, in which a single short wavelength mode is suppressing a long wavelength mode, to the case in which suppression is due to a band of short wavelength modes. The gradient length scale L is now defined by a sum $a_{small}^2/\lambda_{small}$ of the short wavelength modes. We are currently working on further extensions of the definition of L to the late turbulent stage of flow development.

5. Three Dimensional Simulations and Results

We have performed 3D simulations of single-mode nonlinear growth, using the LEEOR3D code [4]. Planar, 3D and cylindrical modes have been compared. We found that the final bubble velocity of all 3D modes is the same, even for a 3D rectangular mode. In Fig. 4 the position of the bubble tip is plotted for the planar, square 3D mode ($k_x = k_y$), rectangular 3D mode ($k_x = 2k_y$) and cylindrical case in which the bubble is on the symmetry axis. It is seen that all 3D bubbles rise similarly, while the planar bubble is slower, consistent with Layzer's prediction for planar and cylindrical bubbles.

The final shape of the single wavelength perturbation was found to take three possible forms [4]: (i) a symmetric checkerboard bubble-spike configuration with saddle points on the diagonals connecting next-nearest-neighbor bubbles or spikes; (ii) bubbles surrounded by a ridge of interconnecting spikes ("bubble-ridge" configuration [5]); or (iii) spikes surrounded by a valley of interconnecting bubbles ("spike-valley" configuration [6]). Which particular late-time configuration is formed depends on the initial conditions that imprint their shape on the late time shape, and the Atwood number. We have found that as the Atwood number rises the saddle-points, which have zero vertical velocity during the linear stage, tend to fall with the spikes (see Fig. 5). Note that when A=0.5 the saddle points are roughly halfway between the bubble and spike penetration, whereas when A=0.9 the saddle point tends to fall with the spikes, essentially creating a circular spike "curtain" around the bubble. This means that at high Atwood ratio a "bubble-ridge" shape will be formed even when the initial perturbation is symmetric with respect to bubbles and spikes (such as a single mode initial perturbation, as is shown in Fig. 5 (b) for A=0.9).

The difference between the above three configurations can explain the different late time shape of the spherical geometry calculations presented by Town & Bell [5] and by Sakagami & Nishihara [6]. We have performed [4] spherical geometry calculations similar to those of Town & Bell and Sakagami & Nishihara. Following Town & Bell we considered a perturbation with dodecahedral symmetry simulating the primary laser induced perturbation in a 12 beam ICF experiment (each beam at the center of a pentagon). Fig. 6 (a) shows the interface between the inner gas and outer layer when the bubbles are in the pentagon centers. Fig. 6 (b) shows the same run but with spikes in the pentagon centers. One can see that the final shape of the interface takes the form of "bubble-ridge" configuration in the case when a bubble is initially present in the pentagon center (Fig. 6 (a)) and a "spike-valley" form when the initial perturbation is inverted (Fig. 6(b)). The fact that in this configuration two different geometries are created is simply an effect of the asymmetry between bubbles and spikes in each of the initial perturbations and should not be related to the spherical geometry of the problem. The same effect can be seen in a two dimensional cylindrical case where a bubble or a spike are initially present on the symmetry axis, or in a three dimensional planar perturbation with hexagonal symmetry.

In contrast with the above case, the case presented by Sakagami & Nishihara consists of a single spherical mode. Their initial conditions consist of a single spherical harmonic mode with $\ell=6$, and m ranging from 0 to 5 and an Atwood number at peak compression of about 0.9. In Fig. 7 (a), we plot our numerical results for a typical full 3D case with $\ell=6$, m=4. One can see that the structure created is approximately that of isolated bubbles surrounded by a ridge of interconnecting spikes ("bubble-ridge"), though there is still some reminder of the initial checkerboard structure. This shape remains when we invert the sign of the initial perturbation. As explained above for the planar case, this fact is a result of the high Atwood number, which is about 0.9 in this case. In Fig 7 (b), we plot for comparison the interface for the case $\ell=6$, m=0, which is an axisymmetric (essentially a 2D) problem. Note the large difference between the three- and two-dimensional axisymmetric structure of the interface. This difference may be important when one is trying to simulate the failure of an ICF target due to laser nonuniformity.

Using the 3D code, we have also calculated the two-mode interaction in cases of moderate wavelength ratios. It has been found that the three dimensional behavior is qualitatively and quantitatively similar to the 2D results. In the case of a short wavelength mode suppressing a long wavelength mode, the amplitude for the onset of suppression (as defined above) is found to be of the form $f(A)=\sqrt{\lambda_1\lambda_2}$ but with $f(A)$ larger than the value in two dimensions by about a factor of 2, qualitatively consistent with the observation that the onset of nonlinear effects in a single RT mode in 3D also occurs at a later stage (or larger amplitude) than in the 2D case.

We have also performed some preliminary simulations of the evolution of random initial perturbations at low-resolution (about 30×30×60). We found the value of α to be about 0.05 (5% to 95% volume fraction of a single fluid), similar to the value found in our 2D simulations of similar and higher resolutions [7]. Youngs found for the high resolution two dimensional case α = 0.04 - 0.05 [8], while for the three dimensional case he found α = 0.05 - 0.06 [9] in the low resolution front tracking simulation and α = 0.03 - 0.04 [10] in the high resolution miscible fluid simulation.

This work was partially supported by the Office of Naval Research and the Department of Energy.

REFERENCES

[1]   D. L. Youngs in "Numerical Methods for Fluid Dynamics", edited by M. J. Baines and K. W. Morton (Academic, New York, 1982).

[2]   D. Ofer, D. Shvarts, Z. Zinamon, S. A. Orszag, Phys. Fluids **B4**, 3549 (1992).

[3]   S.W. Haan, Phys. Fluids **B3**, 2349 (1991).

[4]   J. Hecht, D. Ofer, D. Shvarts, S. A. Orszag, R. L. McCrory, "Analysis of 3D RT Instability", submitted for publication.

[5]   R. P. J. Town, A. R. Bell, Phys. Rev Lett. **67**, 1863 (1991).

[6]   H. Sakagami, K. Nishihara, Phys. Rev. Lett. **65**, 432 (1990).

[7]   N. Freed, D. Ofer, D. Shvarts, S. A. Orszag, Phys. Fluids **A3**, 912 (1991).

[8]   D. L. Youngs, Physica **D12**, 32 (1984)

[9]   D. L. Youngs in "Advances in Compressible Turbulent Mixing", edited by W. P. Dannevik, A. C. Buckingham and C. E. Leith, NTIS Conf-8810234 (1992).

[10]   D. L. Youngs, Phys. Fluids **A3**, 1312 (1991).

Figure Captions

Fig. 1. Suppressing amplitude for various mode pairs and Atwood numbers (A). A positive λ_1/λ_2 ratio indicates that a small λ mode is suppressing a large λ mode. A negative ratio indicates the reverse. The amplitude is normalised by dividing by $\sqrt{\lambda_1\lambda_2}$ in order to judge the accuracy of the rule given in the text. Note that $\ell = 2/\lambda$.

Fig. 2. Illustration of the mechanism whereby a long wavelength mode suppresses the growth of short wavelength "noise" (in this case a single short wavelength). One can see that the long wavelength shear flow tends to pull the short wavelength spikes laterally instead of vertically, resulting in a reduced small scale rms structure. In this case $\ell_{long} = 2$, $\ell_{short} = 30$ and A=0.5. (a) Short wavelength spikes in the presence of a long wavelength shear flow. (b) The same short wavelength perturbation with no long wavelength mode present.

Fig. 3. Suppression of a single long wavelength mode by a band of short wavelength modes. In this case ℓ_{long}=2 and the band consists of $\ell_{short} = 11$-20 with random initial amplitudes and A=0.5. (a) Intermediate times when mode analysis is relevant; (b) late turbulent stage.

Fig. 4. Bubble penetration in 4 single mode cases: square 3D ($k_x = k_y$), rectangular 3D ($k_x = 2k_y$), cylindrical bubble (from the 2D code) and planar 2D. One can easily note that the first three cases, with 3D effects, reach the same bubble velocity whereas the 2D planar case reaches a much lower asymptotic velocity.

Fig. 5. Late stage single mode 3D profile for (a) A=0.5 and (b) A=0.9.

Fig. 6. Spherical perturbations with dodecahedral symmetry. Inversion of the perturbation can result in a: (a) bubble-ridge or (b) spike-valley structure.

Fig. 7. Interface between the inner and outer fluids at late times with single mode spherical RT perturbations with A ~ 0.9: (a) $\ell = 6$, m = 4; (b) $\ell = 6$, m = 0.

Figure 1

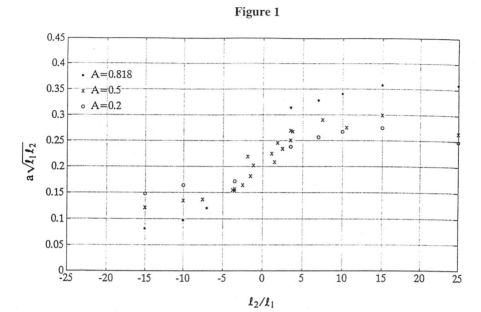

Fig. 1. Suppressing amplitude for various mode pairs and Atwood numbers (A). A positive λ_1/λ_2 ratio indicates that a small λ mode is suppressing a large λ mode. A negative ratio indicates the reverse. The amplitude is normalised by dividing by $\sqrt{\lambda_1\lambda_2}$ in order to judge the accuracy of the rule given in the text. Note that $\ell = 2/\lambda$.

Figure 2(a)

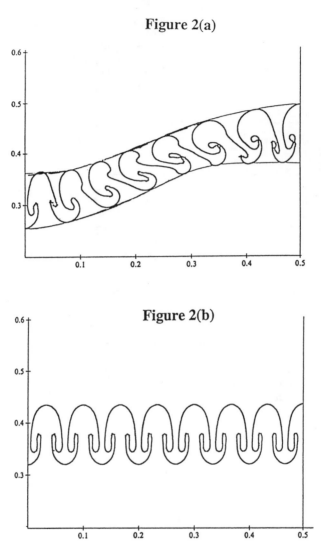

Figure 2(b)

Fig. 2. Illustration of the mechanism whereby a long wavelength mode suppresses the growth of short wavelength "noise" (in this case a single short wavelength). One can see that the long wavelength shear flow tends to pull the short wavelength spikes laterally instead of vertically, resulting in a reduced small scale rms structure. In this case $\ell_{long} = 2$, $\ell_{short} = 30$ and A=0.5. (a) Short wavelength spikes in the presence of a long wavelength shear flow. (b) The same short wavelength perturbation with no long wavelength mode present.

Figure 3(a)

Figure 3(a)   Figure 3(b)

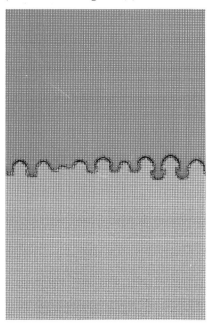

Fig. 3. Suppression of a single long wavelength mode by a band of short wavelength modes. In this case $\ell_{long}=2$ and the band consists of $\ell_{short}=11\text{-}20$ with random initial amplitudes and A=0.5. (a) Intermediate times when mode analysis is relevant; (b) late turbulent stage.

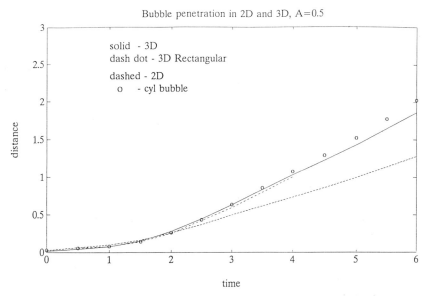

Fig. 4. Bubble penetration in 4 single mode cases: square 3D ($k_X = k_y$), rectangular 3D ($k_X = 2k_y$), cylindrical bubble (from the 2D code) and planar 2D. One can easily note that the first three cases, with 3D effects, reach the same bubble velocity whereas the 2D planar case reaches a much lower asymptotic velocity.

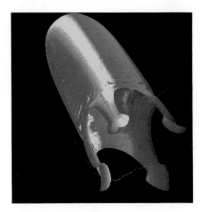

Fig. 5.   Late stage single mode 3D profile for (a) A=0.5 and (b) A=0.9.

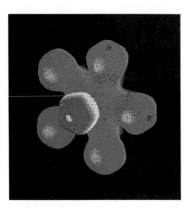

Fig. 6. Spherical perturbations with dodecahedral symmetry. Inversion of the perturbation can result in a: (a) bubble-ridge or (b) spike-valley structure.

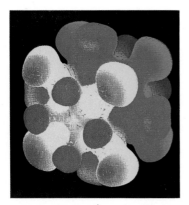
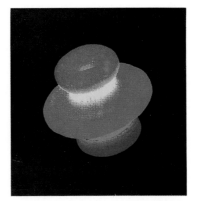

Fig. 7. Interface between the inner and outer fluids at late times with single mode spherical RT perturbations with A ~ 0.9: (a) $\ell = 6$, m = 4; (b) $\ell = 6$, m = 0.

ON THE POSSIBILITY OF A STEADY STATE TOKAMAK

J. M. Dawson, W.J. Nunan, and S. Ma
Physics Department, University of California, Los Angeles CA 90024-1547

TRIBUTE TO TOM STIX

It is a great pleasure for me to speak at this symposium in honor of Tom Stix. I have had the privilege of knowing Tom ever since I started working in plasma physics and fusion at the Princeton Plasma Physics Lab almost 36 years ago. He was a leader of the fusion effort when I arrived and has remained so up to the present time. I vividly remember our interesting discussions on plasma physics. Particularly, I remember how many new and original ideas came from Tom, his ideas for ICRH: the Stix coil, the magnetic beach, and many, many others. Tom not only originated ideas but he built and carried out experiments to test these ideas, as well as many other fundamental concepts in plasma physics. Tom's experiments were always firsts, and many pioneering advances were made by him. Tom's enthusiasm for plasma physics and fusion is infectious; it stimulates and inspires his co-workers and has touched all of Princeton's plasma students. Tom has had a deep interest in teaching plasma physics from the beginning. His excellent course on plasma waves launched many careers. His book on plasma waves, which came from this course, is the standard on the subject, and is an invaluable reference for everyone working in plasma physics. Tom is a generous and caring person which made him an ideal person to lead the Princeton Plasma Physics Graduate Program. It is my great good fortune to have known and worked with Tom, and to have him as a friend.

This symposium is particularly honoring Tom for his guiding of the graduate program in plasma physics at Princeton. For this reason I thought it would be appropriate for me to speak about some work a graduate student of mine, Bill Nunan, is doing, at UCLA. In a real sense the UCLA graduate program in Plasma Physics has many roots in the Princeton program which Tom so skillfully guided.

INTRODUCTION

From the practical and economic point of view a steady state fusion reactor has many advantages. One of the principal arguments for Stellerators over Tokamaks as reactors has been their potential for steady state operation. To achieve steady state for Tokamaks, a good deal of effort has been devoted to a number of methods of non-inductive current drive (RF current drive, energetic beam current drive). At present all of these require of the order of 10 watts of power per amp of current drive. If the total Tokamak current (10 to 20 MA) has to be driven this way, then a very significant fraction of the total power output of the reactor must be devoted to this purpose. This is particularly true when one includes inefficiencies in energy recovery, in RF power production or energetic beam production and losses in delivering this power to the reactor. One must add to this the problems that arise due to the increased heat loads placed on the reactors first wall and other plasma facing components. In addition there are the disadvantages of operating complex equipment and maintaining it in a hostile environment. The loss of efficiency and simplicity substantially detract from the advantages of a steady state for such a current driven Tokamak.

In 1971 the potential for producing a large fraction of the Tokamak current by the so called "Bootstrap" effect was pointed out independently by Kadomtsev and Shafranov[1] and by Bickerton, Connor and Taylor[2]. The reduction in the fraction of the current that must be driven by external sources, due to such Bootstrap current, greatly increases the attractiveness of noninductive current drive.

In the theory of the Bootstrap effect there are two sources of current. First there is the precession in the toroidal direction of the banana orbits of the trapped particles[3]; this is a kind of diamagnetic drift current due to the radial pressure gradient. Second there is a current due to a bias in the direction of flow along field lines of the untapped particles. The second current dominates at start up where it is driven by the ohmic transformer voltage. It can also be driven by the dynamo action of the radial flow (due to transport) of plasma across the poloidal magnetic field.

According to conventional theory, for a steady state, both sources of current disappear on the minor axis of a Tokamak. The banana orbits disappear on the axis and also there is no pressure gradient there. The dynamo effect is largely an MHD one determined by the equation

$$\mathbf{E}_t + \mathbf{v}_r \times \mathbf{B}_p/c = \eta \mathbf{j}_t \qquad (1)$$

Here $\mathbf{E}_t$ is the toroidal electric field and is zero in steady state due to the finite number of volt seconds of the transformer; $\mathbf{v}_r$ is the mean radial velocity of the plasma due to transport; $\mathbf{B}_p$ is the poloidal magnetic field; η is the resistivity and $\mathbf{j}_t$ is the toroidal current density. At the axis $\mathbf{B}_p$ is zero and so is the radial velocity; thus, at the axis, the left hand side of Eq. (1) vanishes and there is nothing to sustain $\mathbf{j}_t$ there.

For these reasons, the theories of Kadomtsev and Shafranov[1], and of Bickerton, Connor and Taylor[2] require a seed current on axis. Existing treatments are quite vague on quantitative values for how large a seed current is required; although, it is certainly a question of critical importance to Tokamak reactors.

The theories of Bootstrap current all start from some analytic approximations to the equations governing current flow (MHD, neoclassical equilibria). These models inevitably leave out physical effects which can play an important role. This is particularly true for steady state situations, where small effects may result in big modifications in regions where other effects vanish. For the short pulse machines that have dominated experiments up to now, the current profiles are controlled by applied voltages, resistivity profiles and magnetic field diffusion. In a steady state machine other mechanisms of current generation and transport can play an important role. In particular these can be important for maintaining the axial seed current where other mechanisms of current drive go to zero.

One important effect which does not appear in fluid theories is the preferential loss, at the outer edge of the plasma, of electrons which are carrying current opposite to that of the main plasma. The situation is illustrated in Fig. 1. The figure shows a cross-section of the torus. The toroidal magnetic field is out of the page as is the plasma current. Consider an electron whose toroidal motion is into the page (its current is out of the page). Such an electron is attracted towards the main plasma current (parallel currents attract) and its banana orbit tends to be inward as shown. On the other hand, an electron whose toroidal motion is out of the page (current into the page) is repelled by the plasma current and its banana orbit tends to be outward as shown. Thus, electrons near the edge carrying current opposite to the current of the main plasma tend to be lost. One method of driving current, which has been proposed, is to inject electrons into the plasma so that their current increases the plasma current; this is one form of what is known as helicity injection. The preferential loss just mentioned accomplishes the same thing; it can be thought of as a loss of helicity of the wrong sign. The numerical experiments that will be discussed later show this effect.

In many of the numerical experiments to be discussed, we have a source of hot plasma at the plasma center (see Fig. 2). The source produces plasma that is not carrying any current. Imagine that an element of this plasma is transported to the outer edge of the plasma by convective motions in the plasma as shown schematically in Fig. 2. Simply moving the plasma across the poloidal field generates current in the desired direction. If when the plasma is near the outer edge it looses electrons carrying current in the wrong direction, then its current is enhanced. Suppose the element of plasma is then convected back to the center of the plasma. This motion, across the poloidal field, drives the current towards zero; if no current had been gained at the edge, the current would be zero when the element arrived back at the center. However, because of the electron loss at the edge, a finite current will remain when it arrives at the center; this current will be in the desired direction and contribute to the required seed current.

In addition to mechanisms for generating current such as the one discussed above, there are mechanisms for transporting current from places where it is generated to places where there is no generation (the convection described above is one such example). There are a number of mechanisms for current transport. First there is classical or neoclassical electron collisional viscosity. This is generally quite small, but to the best of our knowledge, has never been included in any calculations of the Bootstrap current. Second there is transport of electron momentum by waves. There are a large number of waves that can be involved in electron momentum transport; these include electron plasma waves propagating at an angle to the magnetic field, upper hybrid waves, electron cyclotron and Bernstein waves, and electromagnetic radiation such as synchrotron radiation to name a few.

Several years ago Viktor Decyk[4] carried out computer simulations at UCLA showing that one can get wave transport of current without an accompanying diffusion of electrons. He used a two and one half dimensional particle code (two positions, x & y, and three velocities, v_x, v_y, v_z, are followed for each particle). Only electrostatic particle interactions were used; a uniform static magnetic field in the y direction was included; the plasma was bounded with a finite size in the x direction. He started with the plasma carrying current in its central region; the current was produced by a long flat tail on the distribution function for positive velocities. The particle guiding centers diffuse (due to $\mathbf{E} \times \mathbf{B}$ motions) only in the z direction which is an ignorable coordinate (no variation in z) and so there is no particle transport in x. Nevertheless, an x directed transport of the current was observed which was attributed to enhanced wave activity with y directed phase velocities in the region of the enhanced tail. Decyk's results are shown in Figures 3 & 4. Figure 3 shows the plasma configuration that was considered and the initial velocity distribution that exists in the central (shaded) region. Figure 4 shows the initial and final current profiles (showing definite spreading), the initial and final density profiles for the electrons that were carrying the current at $t = 0$ (showing no diffusion), and the observed wave activity vs. phase velocity in the y direction with a comparison to the initial distribution function in the central region. The enhanced wave activity due to the energetic tail is quite evident and Decyk attributes the current transport to it.

There are a number of other subtle mechanisms for electron momentum transport. An example of one is the following: The resistance of a plasma involves momentum transfer between the electrons and the ions. This transfer takes place locally, i.e., within a Debye length or less of the position of the electron and ion. However, because of the finite sizes of the Larmor and banana orbits, this momentum transfer is spread over a finite region of space. If one considers such collisional effects for electrons, it contributes to their viscosity. Because of the much larger size of the ion orbits, the effect is much larger for them. If an ion's orbit carries it from a region of low electron current to a region of larger electron current, it will pick up momentum from the electrons in the high current region and transfer a fraction (~2%, most of it is transferred to other ions) to the electrons in the region of lower electron current. In a fusion reactor the alpha particles may play a larger roll in this effect. Even though there are relatively few alpha particles, because of their large orbits and because they interact more strongly with the electrons than with the ions, their contribution to this type of effect may be substantial.

Perhaps the most important current transfer mechanism will be associated with anomalous transport. Turbulent convective motion of the plasma can transport current from one region of the plasma to another. In calculating this effect one must include all the effects which may modify the current, such as the dynamo effect associated with plasma crossing field lines. Another mechanism which may be important is the possible existence of magnetic islands and ergotic magnetic field lines. Electrons can follow magnetic field lines from regions where a current is being generated by the Bootstrap effect to regions where there is little or no current generation (say the central region). Such a mechanism for current transport has recently been proposed by Weening and Boozer[5]. They assume that if a hollow current profile develops the current profile will be unstable to tearing modes; tearing modes make the magnetic field chaotic, and result in current transport.

While plasma turbulence and chaotic B fields are not well understood, they are believed to be the cause of anomalous electron heat transport; they should also result in anomalous electron momentum transport. It is reasonable to assume that electron momentum is transported with roughly the same diffusion rate that applies to electron energy transport ($\sim 10^4 \text{cm}^2/\text{sec}$). We will use this later when we make numerical estimates of the importance of current transport in present Tokamaks.

On the experimental side there is essentially no data on such current diffusion. Virtually all discharges up to now have been transitory. There are driving electric fields, large resistivity gradients, long L/R relaxation times, not very precise knowledge of sources and sinks of plasma, and there are only rough measurements of the current profile. Thus, the role of dynamo action can not be determined. However, there is good evidence that the Bootstrap effect exists and is strong. There have been experiments in which a substantial amount of the current has has been attributed to the Bootstrap effect[6,7]; the Bootstrap effect has been attributed with driving up to 65% of the current[8]. In these experiments the OH loop voltage was zero or negative for substantial lengths of time while the current was constant; however, in these cases the possibility of transient readjustment of the current profile or the possibility of such things as beam driven currents could not be ruled out.

There have been experiments by R. Taylor[9] on CCT at UCLA in which he runs the machine at steady state, and generates plasma near the center of the tube by injecting electrons there at about 100 eV. These electrons follow the magnetic field lines around and around the torus and ionize neutral atoms as they go, providing the central source of plasma. In addition to the toroidal magnetic field, Taylor also imposes a vertical magnetic field. Taylor finds that a steady state toroidal current is generated and it is always in the direction so as to oppose the tendency of the plasma to expand in major radius. The direction of this current is independent of how the energetic electrons are injected; that is whether they are injected parallel or anti-parallel to the favored direction of the current. For this experiment the toroidal current is rather modest (~1 kA); however, the toroidal and vertical magnetic fields are also low, ~ 1kG and 5G respectively. The poloidal field of the current is only slightly larger than the vertical field. (The minor radius is 30 cm, giving 7G for this field.) Thus, the magnetic surfaces are just barely closed. We believed that the steady toroidal current, seen in Taylor's experiments, is due to the transport of plasma, in the major radial direction, across the vertical magnetic field. It was these experiments that inspired us to carry out the simulations reported in this article.

There have been some relevant recent experiments, which were carefully analyzed[10] using the best standard codes available for analyzing Tokamaks. These experiments contained beam driven current as well as a Bootstrap generated current; for these experiments the predicted beam driven current was larger than the predicted Bootstrap current. However, the observed current was roughly twice the total predicted current, and therefore, the plasma's self-generated current was at least 3 times what could be accounted for by the conventional Bootstrap theory. Such experiments give a strong indication that additional mechanisms are driving current besides those considered by existing theories.

Given all of the above mechanisms for current transport, none of which has been explored in detail, given our lack of understanding of turbulence and turbulent transport in Tokamaks, and given the lack of precise experimental data, it is impossible to make any accurate predictions of the magnitude of the current diffusion rate. For this reason we started to explore the question of current generation in Tokamaks using numerical modeling. Details of the numerical model are given in the next section. We also backed up our model results with a simple semi-empirical analytic theory. This theory involves an effective current transport through a modeled electron viscosity. The theory involves an electron viscosity coefficient; for relating our results to real devices, a reasonable working hypothesis is to choose this viscosity so that the electron momentum diffusion rate is equal to the energy diffusion rate.

COMPUTER MODELING AND RESULTS

The Model

As mentioned above we have been exploring Bootstrap current generation in Tokamaks using computer modeling. The model used was a 2-1/2 dimensional (2 position variables and 3 velocities) fully electromagnetic PIC model[11,12]. The geometry is what we call a quasi-torus; that is we straighten out the torus but keep a gradient of the axial magnetic field, (z directed, B_z) in the x direction; x corresponds to the major radius. We take the y direction to be parallel to the major axis of the torus. Thus, gradient B drifts are retained but there is no curvature drift due to motion around the torus (in the z direction.) The model used in the calculation is shown in Figs. 5 & 6.

It is possible to use a uniform B_z; this corresponds to an infinite major radius torus or a straight cylinder. We have done simulations with a straight cylinder to compare with those of a quasi-torus. We impose a small B_y field for quasi-toroidal equilibrium; however, for the case of a straight cylinder no such field is required or included.

The particles move in the x and y directions and are allowed to have a z component of their velocity, so that there can be an axial current. However, all quantities are independent of z. The full electromagnetic fields of the particles are computed and the particles move in these self consistent fields plus the imposed B_z and B_y fields mentioned above. The combination of the poloidal B field due to the axial current and the nonuniform B_z field produces both banana and passing orbits for the particles just as in real Tokamaks.

We follow both the electron and ion dynamics; this insures that the model is a real dynamical system and any processes it exhibits represent physical processes that can take place. Because we must follow the the plasma on an electron time scale, small ion to electron mass ratios were used in the simulations (1 to 1 up to 9 to 1). Typical simulations used 50,000 electrons and ions, the sizes of the systems varied from 16 to 64 c/ω_p across, in terms of Larmor radii the systems varied from 64 to 256 (64 to 128) electron (ion) Larmor radii across, and runs up to $\omega_p t \sim 50,000$ were made.

The walls of the torus are square and are perfect conductors. A circular limiter is employed; this limiter is tangent to the conducting boundaries of the tube. Whenever a particle goes outside the limiter it is removed from the system. The particles are then re-injected as neutral pairs, so as to maintain the plasma. The re-injected particles have the same temperature as the initial plasma; that is they have Maxwellian distributions in v_x, v_y & v_z. The temperatures are the same for all three velocities, and the electron and ion temperatures are equal. The average z velocity of the injected electrons (ions) is zero, so that no current is injected. For most of our calculations the particles are re-injected into the central region of the plasma. However, runs have been made in which they are injected into a circular annulus to simulate edge fueling, as in gas puffing.

The grad $\mathbf{B}_z$ drifts tend to polarize the plasma and cause it to be expelled outward (in the x direction). This expulsion of the plasma is prevented by imposing a vertical $\mathbf{B}_y$ field (in the y direction) and an axial current, so that the expulsion force is just balanced by the $\mathbf{j} \times \mathbf{B}$ force. We start the plasma out very nearly in an equilibrium state (force balance). Since it is very hard to do this precisely there is initially some rapid transient motion of the plasma. However, following this early phase the plasma settles down and executes only small oscillations about the equilibrium position. There is a steady loss of particles at the limiter. These particles are re-injected at the desired place in the plasma as neutral pairs with a thermal distribution at the initial temperature. We keep track of the number of particles lost and re-injected, contours of the electrostatic potential, contours of the current density, the total plasma current, and a number of other quantities.

Results of the Simulations

Figures 7 through 13 show typical results from simulations. The plasma consisted of 50,000 electrons and ions with an ion to electron mass ratio of 9. The system was 128 grid points across (all distances are measured in grid spacings) which corresponded to $\sim 26 \ c/\omega_p$. The toroidal magnetic field at the center gave a ratio of the electron cyclotron frequency to the plasma frequency of $\omega_c/\omega_p = 2$; the electron Larmor radius was $\rho_e = 0.5$ and the Debye length was $\lambda_D = 1$. The scale length for the variation of the z magnetic field was L = 256 which corresponded to an aspect ratio for the torus of 4. The vertical magnetic field, $\mathbf{B}_y$, was 0.875 x 10^{-3} of the toroidal field and the magnetic field of the axial current at the edge of the plasma was 0.1 of the toridal field. The plasma ß associated with the toridal field was $ß_t = 0.04$ and associated with the poloidal field was $ß_p = 4$.

Figure 7 shows the number of ions and electrons lost vs. time and the time variation of the total I_z. There is a steady loss of particles; however, the rate of loss does decrease as time goes on. The magnitude of the observed loss is of the order of that which would be predicted from the convective diffusion observed by Dawson and Okuda[13,14] for a two dimensional simulation plasma. (Diffusion coefficient D = $[2\pi]^{-1/2}[v_{ti}/n^{1/2}] [\ln\{L/\rho_i\}]$; v_{ti} is the ion thermal velocity; n is the two dimensional particle density; L is the system size; and ρ_i is the ion Larmor radius.) The scaling seems to differ somewhat from this formula, but the magnitude is about right. For this reason we believe that the plasma loss seen here is due to convective cell transport. This belief is strengthened by plots of equipotential contours shown in Fig. 8. This "anomalous" loss is much faster than collisional diffusion. Collisional diffusion loss times are $\sim (L/2\rho_e)^2\nu_{ei}$, which in this case is $\omega_p t \sim 10^7$ or about 2 orders of magnitude longer than the observed loss time. While the loss seen here is probably different from that seen in real Tokamaks, we believe that it may have much in common with that process (i.e. both are due to convective transport).

Figure 7 also shows the time dependence of the total z current. Initially this current is chosen so that the plasma is close to equilibrium, i.e., in force balance in the x direction,

$$\mathbf{j}_z \times B_y/c = (nm_i v_{ti}^2) \nabla \ln(B) \ . \qquad (2)$$

Because of the self consistent fields generated by the plasma, because Eq. (2) is a lumped circuit model of the plasma, and because in actuality we must have force balance at each point in the plasma to have equilibrium, it is difficult to precisely start in equilibrium. As a result the initial plasma deviates slightly from equilibrium and there is some rapid initial adjustment of the plasma position and size which results in the initial transient oscillations that are seen in Fig. 7. However, after this initial rapid adjustment, $\mathbf{j}_z$ rises almost linearly through out the rest of the run; at the end of the run it is about 20% larger than it was at the start. The current shows no signs of saturation in the runs we have made to date.

Figure 8 shows plots of equi-potential contours about half way through the run; this plot is typical of the potential contours at any time; of course the details are continually changing with time. Also shown in this Figure are the potential variation along a vertical slice at x = 70. From both plots we see that there is a steep sheath region at the edge, and the center of the plasma is negative. This is caused by the preferential loss of ions at the limiter because of their large banana orbits; this is in agreement with the theory of Shaing[15]. The negative plasma potential inhibits the loss of ions and enhances the loss of electrons; the plasma charges up until the ion loss equals the electron loss. Inside the plasma the potential contours show a structure with many peaks and valleys. To the extent that the particles move with an $\mathbf{E} \times \mathbf{B}$ motion these contours give the orbits of particles; the motions are vortices or convective cells. This convective motion gives the diffusion; more details can be found in references 13 and 14.

Figure 9 shows contours of $\mathbf{j}_z$ as well as a plot of $\mathbf{j}_z$ taken along a vertical slice at x = 65. The plot is made for a time roughly half way through the run. This plot is typical of plots made at any time during the run; of course the details change with time. The plot shows that the current consists of a large number of filaments; they pretty much have the same direction (sign). We believe that this structure has the following cause. The vortex motion of the plasma causes it to flow across magnetic surfaces. The motion across a magnetic surface tends to generate $\mathbf{j}_z$ through the $\mathbf{v} \times \mathbf{B}/c$ effective electric field. Such motion will generate current in the positive direction on one side of the vortex and in the negative direction on the other. However, since there is more material flowing out than in (because of the central source) there is a net tendency to drive current in the direction it is already flowing. Thus, the outward flow tends to amplify the existing current. One can also think of this in terms of conservation of canonical momentum; since all quantities are independent of z, the canonical z momentum, $\mathbf{p}_z = mv_z + qA_z/c$, of a particle is a constant of the motion (until the particle is removed from the system at the edge). As the particles move out their $\mathbf{v}_z$

changes to keep $\mathbf{p}_z$ constant. The outward flow of the plasma is such as to increase the current in the plasma.

Another way to think of the current filament formation is the following. The ions and electrons have different motions due their grad $\mathbf{B}$ drifts, and because their motions in the vortices are different due to finite Larmor radius effects, and due to their different rates of response to what a moving fluid element sees as time dependent $\mathbf{E}$ & $\mathbf{B}$ fields (polarization drifts). As a result charge separation arises, and the electrons move along $\mathbf{B}$ (nearly parallel to z) to try to neutralize these charge differences. This and the dynamo picture are two different ways to think about the same phenomenon. (Of course all the same physics must be included in each picture to get them to agree.)

Despite the filamentary nature of the current profile, the magnetic surfaces are pretty well behaved. Figure 10 shows a plot of the magnetic surfaces projected on the x, y plane (surfaces of constant A_z) at the same time as the current profile shown in Figure 9. This is probably not too surprising since these curves are approximately the double integral of the current density ($-\nabla^2 A_z \approx 4\pi \mathbf{j}_z/c$), since the displacement current is small.

One advantage that computer modeling has over laboratory experiments is the possibility to make as detailed measurements as we desire of quantities of interest. One quantity of interest is the z velocity distribution of the lost particles. From our earlier discussion, we expect that particles near the edge, carrying current opposite to the current of the main plasma will be preferentially lost. We can check this prediction by measuring the distribution function of z velocities at birth ($f_b[v_z]$) for the lost particles. Figure 11 shows such distribution functions obtained from all the lost particles up to the time of the measurement, for both electrons (upper left figure) and ions (upper right figure). We see that the peak of the distribution is shifted to the positive side of zero for the electrons, and in the negative direction for ions. This shift is just the opposite of what would occur if the particles were removed from the bulk plasma at random, and hence had a drift corresponding to the average current. Thus our conjecture of preferential loss of particles near the edge which are carrying current opposite to the plasma is born out.

It is possible to test our understanding even further. Since we expect the canonical z momentum of the particles to be conserved, we should be able to predict the birth z velocity distribution ($f_c[v_z]$) of the lost electrons from a knowledge of their final z velocities and their values of A_z at the places where they are born and lost. The lower graphs in Fig. 11 show the results of such predictions; the agreement with the upper graphs is excellent.

We see from the above that, to a large extent, the current drive arises from plasma crossing poloidal flux. Thus, fueling at the center should be most effective. Edge fueling, such as provided by gas puffing or recycling, should be much less effective. Only the small amount of magnetic flux the neutrals cross before they are ionized would contribute to current drive. However, if our picture of the preferential loss of edge electrons carrying current the wrong way is correct, there should still be some current generation. It is possible to test this idea using the model. Instead of re-injecting plasma at the center we can re-inject it in a narrow annulus at the edge. We have carried out a simulation in which we did that; the current dependence for this run is shown in Figure 12. We see from this figure that there is still current generation with edge fueling. The current rise is much weaker than for the central fueling case. Also, if we look at the loss (and hence the fueling rate) shown in Figure 12 we see that it is about 20 times larger than for central fueling, so that the current generated per injected pair is one to two orders of magnitude smaller. This result also indicates that the ideas concerning preferential loss of the appropriate electrons is correct, and that central fueling is much more efficient for current generation. In a real Tokamak central fueling may be difficult. What should really count as far as current drive goes is the amount of poloidal flux outside the point of deposition. If injected pellets can penetrate 30% of the way to the axis, then they would cross approximately half the poloidal flux, and current generation should be efficient.

The standard theory of the Bootstrap current involves neoclassical effects in a torus. The picture that the current is generated by radial flow across the poloidal magnetic field would not require neoclassical effects; it is of interest to test whether or not they are needed. For this reason we carried out numerical experiments on a straight cylinder. For these simulations the imposed B_z was taken to be constant across the cross-section and no B_y was imposed. Central fueling was used for this simulation. Figure 13 shows a plot of the current vs. time for this simulation. The current again starts to rise very much as in the centrally fueled, toroidal case shown in Fig. 7. The rate of rise of the current is a little smaller in this case; however, it is within a factor of 2 of the quasi-torus case. Thus, the effect is not basically a neoclassical one. That this should be true seems obvious from the conservation of p_z arguments, $p_z = mv_z + qA_z/c$.

ANALYTICAL MODEL

In view of the computer simulation results, it is clear that there are physical processes taking place in Tokamak plasmas that are not included in the standard theories of Bootstrap current drive. At the beginning of this article, we discussed a variety of current transport mechanisms which could play a role in Bootstrap current drive, and in particular in providing the seed current required by the standard theories. We have developed an analytic treatment which incorporates an empirical electron viscosity to model current transport. The model is similar to the one used in the work of Weening and Boozer[5]. Since we have seen that current growth exists in straight geometry as well as toroidal geometry, we expect that it should also show up in slab geometry. Since our aim is to gain insight into self generated currents rather than try to make precise predictions, (and we probably don't understand current transport well enough to make such predictions), we will use a simple slab model for our analysis. The model is shown in Fig. 14; the slab is parallel to the y, z plane. Plasma is injected into the center of the plasma, x = 0, and flows outward to the boundaries, x = $\pm$ a, where it is absorbed. We assume that it flows outward at a velocity $\pm v_x$; for simplicity we take the plasma density to be constant, so that v_x is constant by the continuity equation. We assume that the plasma carries a current in the z direction; we apply just enough E_z to maintain the current at the desired value. This condition is equivalent to specifying B_y at the plasma boundaries (x = $\pm$a). By symmetry, the values of $v_x(-x)$,

$B_y(-x)$, $E_z(-x)$, $j_z(-x)$ are given by $-v_x(x)$, $-B_y(x)$, $E_z(x)$, $j_z(x)$ respectively. A component of B in the z direction is allowed, but it plays no role in the calculations.

In the model we modify Ohm's Law to include an empirical electron viscosity:

$$E + v \times B/c = \eta \cdot j - (\nabla \cdot \mu \cdot \nabla j)m_e/(e^2 n) , \qquad (3)$$

where μ is the electron momentum diffusion coefficient (specific viscosity). Because of the slab geometry, we can ignore the vector character of v, B, E, j; we will also treat η and μ as scalars rather than tensors; we also treat them as constants. Using $4\pi j_z = c\partial B_y/\partial x$ and $\eta = 4\pi v_{ei}/\omega_p^2$ (where v_{ei} is the electron-ion collision frequency), Equation (3) becomes

$$E_z + v_x B_y/c = \frac{c}{\omega_p^2}\left[v_{ei} \frac{\partial B_y}{\partial x} - \mu \frac{\partial^3 B_y}{\partial x^3}\right] . \qquad (4)$$

If we scale x to the size of the plasma $\xi = x/a$, and we divide equation (4) by $v_{ei}c/(a\omega_p^2)$, we can write this equation in terms of normalized viscosity, $\Gamma = \mu/(v_{ei}a^2)$, as follows:

$$\tilde{E}_z + \tilde{v}_x B_y = \partial B_y/\partial \xi - \Gamma \partial^3 B_y/\partial \xi^3 . \qquad (5)$$

$\tilde{E}_z = E_z a\omega_p^2/(cv_{ei})$ and $\tilde{v}_x = v_x a\omega_p^2/(c^2 v_{ei})$ which is the magnetic Reynolds number. The boundary conditions that we impose are

$$B_y = 0 \text{ at } \xi = 0; \ \partial^2 B_y/\partial \xi^2 = 0 \text{ at } \xi = 0; \ \partial^2 B_y/\partial \xi^2 = 0 \text{ at } \xi = 1 . \qquad (6)$$

In addition we require that at $\xi = \pm 1$, B_y have the value consistent with the total current in the slab. This is needed to determine $\tilde{E}_z$ which is a free parameter. The second of these boundary conditions assumes that the viscous force on the electrons must vanish at the boundary of the plasma. Since we have argued that there can be a preferential loss of electrons moving parallel to the current, this amounts to a momentum transfer between the electrons and the limiter. If we wish to include this effect, we must alter the second boundary condition to be consistent with this momentum transfer. We have carried out some calculations of this type; they show some quantitative differences, but qualitatively they are the same.

We have solved equation (5) for different values of the parameter Γ. Figures 15 to 17 show some results of these calculations. Figure 15 is a plot of the $\tilde{E}_z$ required to maintain the given current vs. the magnetic Reynolds number; curves for various values of Γ are shown. For large electron viscosity (large Γ's) all the electrons must move with the same z velocity; current generated in the outer portions of the plasma drags along the central electrons providing the seed current. For large values of Γ, $\tilde{E}_z$ passes through zero at $\tilde{v}_x = 2$; thus, no driving electric field is required to maintain the current. For larger values of $\tilde{v}_x$ we can run the transformer in reverse and extract power from the system; the source of the energy is, of course, the injected hot plasma. As the value of Γ is decreased, the value of $\tilde{v}_x$ at which $\tilde{E}_z$ passes through zero moves to larger and larger values.

Figure 16 shows plots of $\tilde{E}_z$ vs. $\tilde{v}_x$ for $\Gamma = 0$ and 5×10^{-4}. (Note that the lowest value on the v_x axis is 15, not 0). For $\Gamma = 5 \times 10^{-4}$, $\tilde{E}_z$ becomes zero at $\tilde{v}_x \approx 17$. Even if the flow velocity is lower (e.g., $\tilde{v}_x = 15$), the electric field required to maintain the current is considerably less than would be the case if the viscosity were entirely absent ($\Gamma = 0$).

From our calculations, we can determine the dependence of the magnetic Reynolds number at $\tilde{E}_z = 0$ on normalized viscosity, Γ. Figure 17 shows a plot of $\tilde{v}_x$ vs. $\Gamma^{-1/2}$. The dependence is essentially linear, and the regression fit is shown at the top of the graph.

To explore what the situation might be for an existing Tokamak, we must make some reasonable estimate of Γ. To do this we will assume that the electron momentum diffusion coefficient is equal to the electron energy diffusion coefficient; we take this to be 10^4 cm^2/sec. For an electron temperature of 3 keV, and an electron density of 3×10^{13}/cm^3 with an effective Z of 2, then $v_{ei} \approx 17600$/s. If we take the minor radius of the machine to be one meter, then Γ comes out to be 5.7×10^{-5}. From the regression in Fig. 17 we see that the $\tilde{v}_x$ required for steady state is 49. Taking the particle confinement time to be 1 sec., we get a magnetic Reynolds number of $\tilde{v}_x = 60$. Thus, the Reynolds number is greater than the value that this model requires to sustain the current. At smaller electron temperatures, the assumed flow velocity of 100 cm/s would be inadequate to sustain the current, but at higher electron temperatures, that velocity would be much more than sufficient. Given the crudeness of the calculation, which neglects the effects of spatial profiles of temperature, density, and fueling, we can not say whether or not present Tokamaks could sustain their current if they were centrally fuelled and properly heated. More detailed calculations could be made using the correct geometry and a range of electron specific viscosities around 10^4 cm^2/sec. These would give us a better estimate of whether or not present Tokamaks could exhibit a steady state current; however, given our lack of knowledge of current transport, there would still be large uncertainties. It is clear that this is an important question for practical Tokamak reactors. Physics experiments, theory, and computer simulations that address the question of current diffusion, and the degree to which long pulsed or steady state Tokamaks can sustain their own current when centrally fueled and heated are certainly called for. The importance of achieving steady state operation of a Tokamak reactor without complex current drive equipment can hardly be over stated.

ACKNOWLEDGEMENTS

The authors would like to acknowledge valuable discussions with Viktor Decyk. This work was supported by USDOE and NSF.

REFERENCES

1. Kadomtsev, B.B. and V.D. Shafranov, "A Stationary Tokamak", Proc. 4th Int. Conf. on Plasma Physics and Controlled Nuclear Fusion Research, Paper IAEA-CN-28/F-10, Nuclear Fusion Supplement, p. 209 (Madison, WI, 1971).

2. Bickerton, R.J., J.W. Connor, and J.B. Taylor, "Diffusion Driven Plasma Currents and Bootstrap Tokamaks", Nature Physical Science, Vol. 229, p.110 (Jan. 25, 1971).

3. Hazeltine, R.D. "Review of Neoclassical Transport Theory", Advances in Plasma Physics, Vol. 6, Part II. Simon and Thompson (Editors), p. 273-309, Interscience, New York, London, Sydney, Toronto (1976).

4. Decyk, V.K., G.J. Morales, J.M. Dawson, and H. Abe, "Radial Diffusion of Plasma Current Due to Secondary Emission of Electrostatic Waves by Tail Electrons", Proc. of 13th European Conf. on Controlled Fusion and Plasma Heating, Schliersee, FRG, April, 1986, Europhysics Conference Abstracts, Vol. 10C, Part II, pp 358-361.

5. Weening, R.H. and A.H. Boozer, "Completely Bootstrapped Tokamak," Phys. Fluids B, 4, (1), 159 (1992).

6. Zarnstorff, M.C., et. al., "Bootstrap Current in TFTR", Phys. Rev. Let., 60 (13) 1306 (1988).

7. O'Brien, D.P., C.D. Challis, J.G. Cordey, J.J. Ellis, G. Jackson, L.L. Lao, P.M. Stubberfield, and T.S. Taylor, "Review of Recent Advances and Current Drive on Textor," Proc. of the 18th EPS Conference on Controlled Fusion and Plasma Physics, Berlin, FRG, June 3-7, 1991.

8. Sabbagh, S.A., et. al., "High Poloidal Beta Equilibria in the Tokamak Fusion Test Reactor Limited by a Natural Inboard Poloidal Field Null," Phys. Fluids B, 3 (8), 2277 (1991).

9. Taylor, R.J., M. Brown, J. Evens, K. Lai, J. Liberati, T.K. Mau, S. Park, and P. Pribyl, "Transport Induced Current Drive In Tokamaks", Bulletin of the APS, 32 (9), abstract 2Q9, 1747, (1987).

10. Messiaen, A.M., et. al., "Review of Recent Advances in Heating and Current Drive on Textor", Europhysics Topical Conference on Radiofrequency Heating and Current Drive of Fusion Devices, Laboratory for Plasma Physics, Ecole Royale Militaire-Koninklijke Militaire School, Brussels, 7-10 July, 1992. To Appear in Plasma Physics and Controlled Fusion.

11. Dawson, J.M., "Particle Simulation of Plasmas", Rev. of Mod. Phys., 55, 403, (1983).

12. Birdsall, C.K., and A.B. Langdon, "Plasma Physics via Computer Simulation", McGraw-Hill, New York, (1985).

13. Dawson, J.M., H. Okuda, and R.N. Carlile, "Numerical Simulation of Plasma Diffusion Across a Magnetic Field in Two Dimensions", Phys. Rev. Lett., 27, 419, (1971).

14. Okuda, H. and J.M. Dawson, "Theory and Numerical Simulation on Plasma Diffusion Across a Magnetic Field", Phys. Fluids, 16, 408, (1973).

15. Shaing, K.C., "Ion Orbit Loss and Poloidal Plasma Rotation in Tokamaks," Phys. Fluids B 4 (1), 171 (1992).

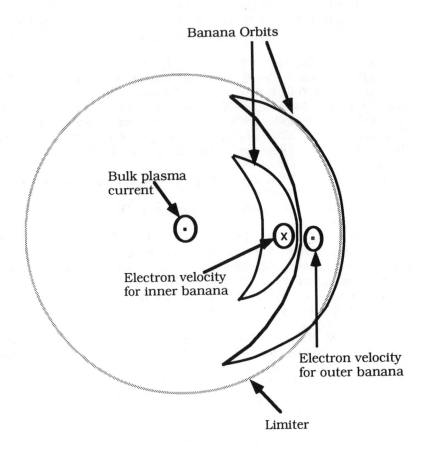

Figure 1 Preferential loss of electrons with current opposite the plasma current.

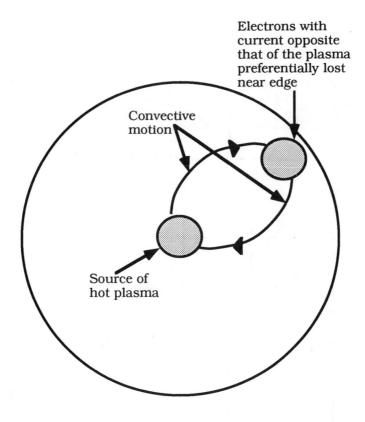

Figure 2 Central source of plasma and its convective transport to the plasma edge and back.

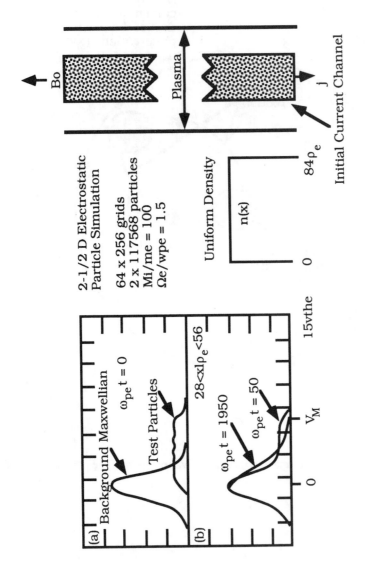

Figure 3(a) Initial parallel velocity distribution of electrons inside the current channel consists of a background Maxwellian plus a uniformly distributed ($0 < V_\parallel < V_m$) set of test particles; (b) Evolution of parallel electron velocity distribution inside current channel shows relaxation of initial "slideaway" distribution into a half-Maxwellian for $V_\parallel > 0$, and no change from the initial Maxwellian for $V_\parallel < 0$.

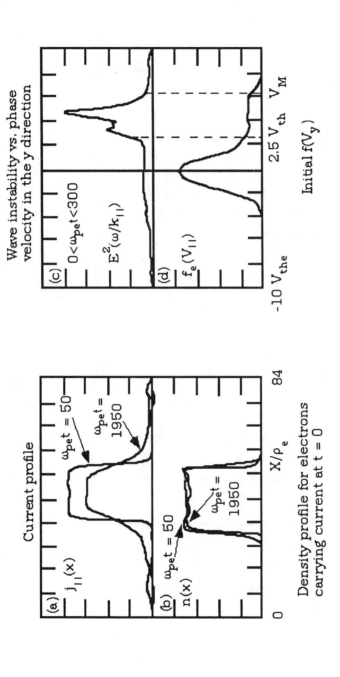

Figure 4(a) Initial and final parallel current profile shows substantial widening of current channel across the magnetic field; (b) Initial and final test particle density shows virtually no diffusion of test particles across the magnetic field; (c) Electric field fluctuation energy versus parallel phase velocity at early times (d) shows enhancement for velocities corresponding to the non-Maxwellian part of the velocity distribution.

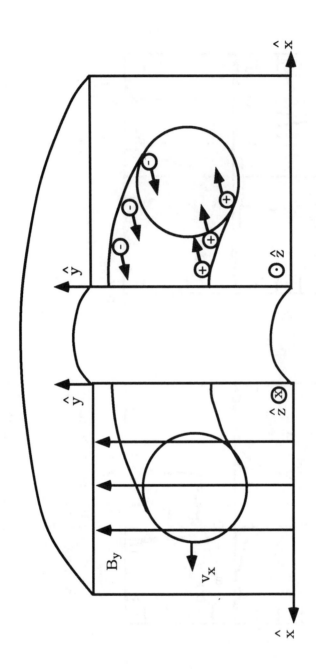

Figure 5   Toroidal System Modelled. A Maxwellian plasma in D.C. toroidal and vertical fields spontaneously develops a large toroidal current. ∇B drift causes charge separation; B_y gives a "slope" to the total B field lines; Particles flow along $\vec{B}$ to neutralize charge; $B_z \gg B_y$, therefore $J_z \gg J_y$.

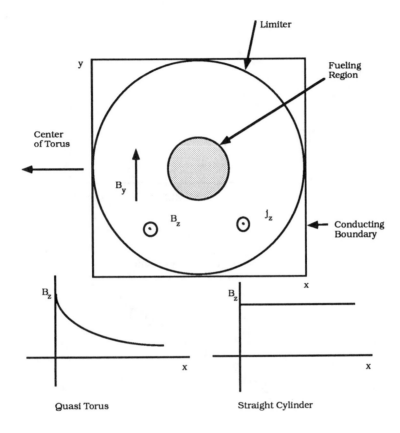

Figure 6   Quasi Torus model used in the simulations.

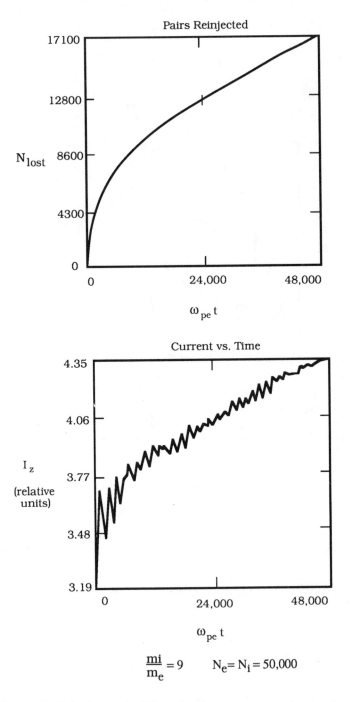

Figure 7   Particles lost and reinjected and current generation vs. time.

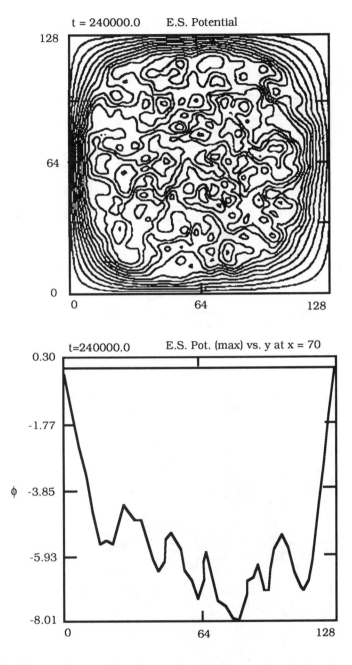

Figure 8 Top, equipotential contours about half way through the run; Bottom, the potential variation along vertical line at x = 70.

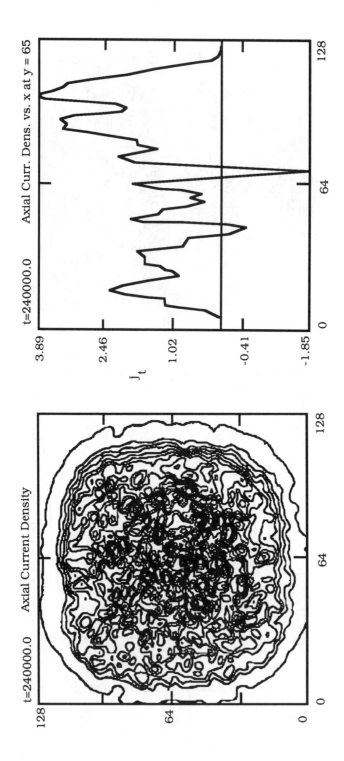

Figure 9 (a) contours of a j_z about half way through the run; (b) the j_z current profile along a vertical slice at x = 65.

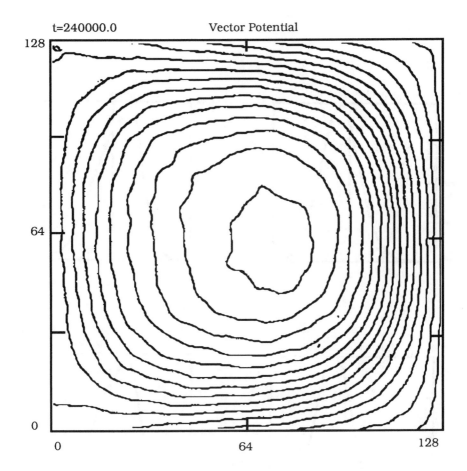

Figure 10 Plot of magnetic surfaces (surfaces of constant A_z).

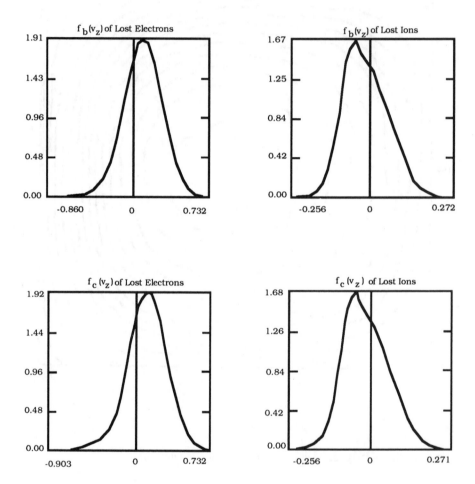

Figure 11 Upper left, z velocity distribution of lost electrons; Upper right, z velocity distribution of lost ions; Lower left, calculated z velocity distribution of lost electrons using conservation of canonical z momentum; Lower right, calculated z velocity distribution of lost ions using conservation of canonical z momentum.

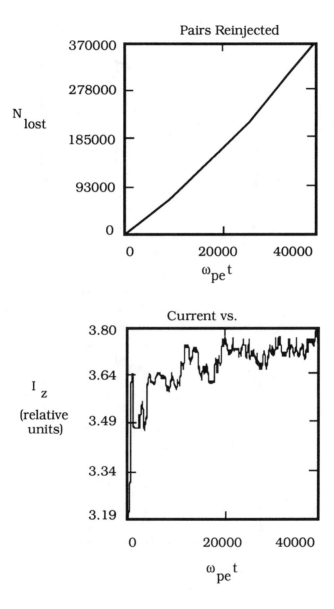

Figure 12 Particles lost and reinjected and current generation vs. time for edge fueling.

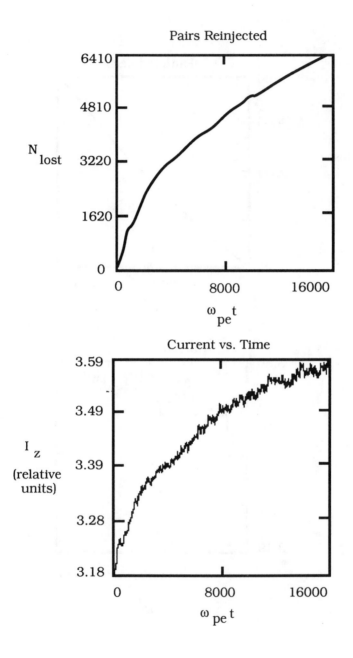

Figure 13   Particles lost and reinjected and current generation vs. time in a straight cylinder with central fueling.

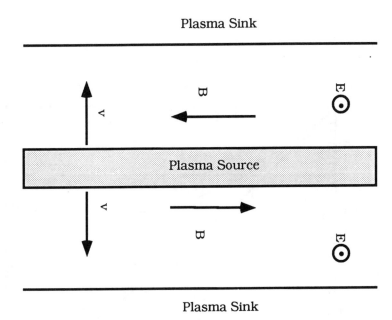

Figure 14   Analytic model used for bootstrap current calculation.

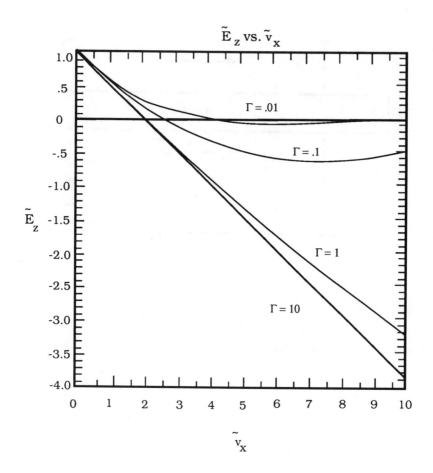

Figure 15 $\tilde{E}_z$ vs. $\tilde{v}_x$ (the magnetic Reynolds number) for various values of Γ.

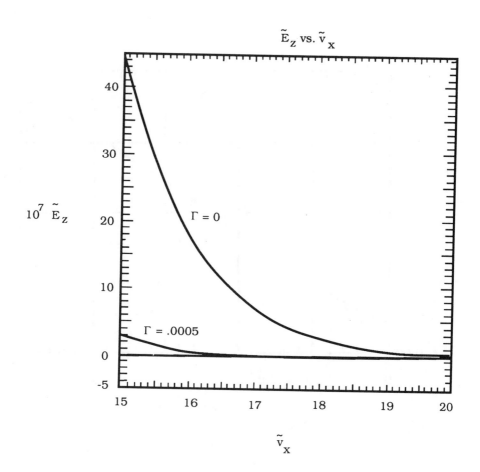

Figure 16   Plots of $\tilde{E}_z$ vs. $\tilde{v}_x$ for $\Gamma = 0$ and 5×10^{-4}.

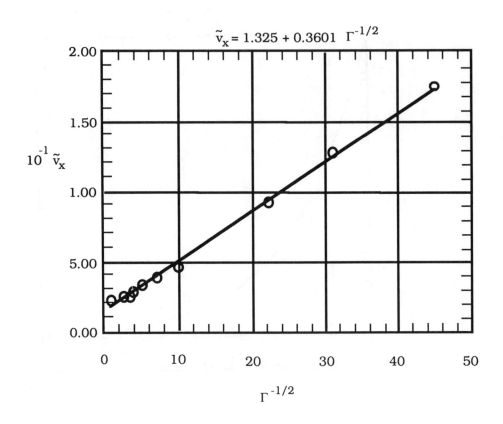

Figure 17   Plot of $\tilde{v}_x$ at which $\tilde{E}_z = 0$ vs. $\Gamma^{-1/2}$.

RESEARCH ON MITIGATION OF STRATOSPHERIC OZONE DEPLETION

A. Y. Wong
Department of Physics
University of California, Los Angeles
Los Angeles, California 90024-1547
(E-Mail: awong@physics.ucla.edu)

ABSTRACT

Chlorine atoms released from CFCs by solar Ultraviolet (UV) radiation and from natural sources, are effective catalytic agents for the destruction of stratospheric ozone. Research into large-scale mitigation methods is based on charging the chlorine radical, converting it into negative ions of low reactivity. Generation of charges and subsequent removal of chlorine ions by atmospheric platforms and electromagnetic waves are described. This method is generally applicable to all halogens. This research is guided by the principle that the solution should be as non-intrusive environmentally as possible; i.e. no chemicals are to be injected. The large-scale mitigation requires the process to be energy efficient and to utilize energy sources already present in the atmosphere. Because of the wide variety of remediation concepts, each is being tested using a combination of laboratory and field experiments together with computer modeling. The first laboratory demonstration of ozone depletion and subsequent recovery due to charge injection is presented.

INTRODUCTION

This paper was given in the symposium, May 19, 1992 in honor of Professor Tom Stix who has made various notable contributions to the field of science. Among his many interests two are being touched upon here - the depletion of stratospheric ozone and ion cyclotron waves.

The existence of an Antarctica ozone hole has been firmly established by satellites and ground measurements. While there are still debates about their origins, chlorine atoms are destroying ozone molecules catalytically with a large multiplicative factor of the order of 10^5. The large reservoir of chlorofluorocarbons (CFCs) in the atmosphere and its potential for ozone destruction for many decades raises the necessity of searching for mitigation measures.[1,2,3,4] The cost of replacing CFCs in present systems and the lack of a perfect substitute mean that the release of chlorine-containing chemicals will continue. Any conservation effort on earth, no matter how complete, will not be felt for decades to come. The convection of the heavier CFC molecules is a slow process and takes many years to reach the stratosphere.

It is now well established by laboratory experiments and theoretical calculations that chlorine radicals destroy ozone molecules very effectively by catalytic reactions. These chlorine atoms are products of solar UV radiation, dissociating CFC molecules in the stratosphere. Chlorine atoms belong to the halogen group which is very reactive because of its lack of one electron in its outer shell. However once its outer shell is filled by acquiring one electron to become a negative ion, it becomes inert and its reaction rate with ozone is reduced by orders of magnitude. Our proposed research is therefore to search for the most effective conversion of chlorine atoms into negative ions and subsequently collect them. A Middle Atmosphere Platform (MAP) using helium for floatation and solar energy for its power is designed to provide metallic surfaces from which electrons can be released by solar radiation. Negative ions are formed through electron attachment instead of ionization in order to conserve energy. These negative ions are collected at positively-biased electrodes, resulting in its permanent removal. The resulting dc current between electrodes provides a continuous flow of electrons and negative charges without causing charge imbalance in the affected volume. This Middle Atmosphere Platform can also be effectively used to charge and subsequently collect particulates in the Polar Stratospheric Clouds (PSC), which are reservoirs of Cl radicals. The concentrations of chlorine reservoirs in clouds will make the entire process much more efficient by reducing the volume of collection.

An additional method of removal of charged pollutants such as chlorine atoms, which are the minority species in the upper atmosphere, is described. In the polar regions charged species can be transported upward by convection from the warmer stratosphere to the mesosphere. Negatively charged species will drift upward because of the downward pointing fair-weather electric fields, which can be artificially enhanced by ionospheric heating. When ions reach the transition height (above 100 km) between the collisional and collisionless regimes, they can be resonantly accelerated by EM fields, oscillating at their ion cyclotron frequency. These low frequency EM fields of large spatial extents (30-100 km) have been successfully excited in the upper atmosphere by high power HF electromagnetic waves modulating the auroral electrojet.

We wish to emphasize that our methods of charge injection are distinctly different from other authors[5] who suggest injecting large quantities of propane into the atmosphere. These chemicals, once released, cannot be easily recaptured. Furthermore propane as a green-house gas will cause other problems such as global warming. We favor the injection of charges because they are controllable and recoverable.

In this paper we will discuss the principle[1,2] of converting Cl into negative ions to slow down its reaction rate with ozone in section I. Laboratory experiments and computer modeling used to demonstrate the catalytic destruction and the effects of charging ions by attachment are described in section II. Methods of charging the atmosphere, to produce ions by attachment are discussed in section III. Low energy electrons of 1eV or less are preferred because of their large attachment probability, low cost of production and their inability to dissociate ozone. The collection of ions and charged particulates is also considered in this section. A Middle Atmosphere Platform (MAP) using solar energy for charging and subsequent collection of charged pollutants is presented in section IV. The use of solar energy and the minimization of ground-based power sources are important considerations in a large scale collection strategy. In section V a large-scale method of selective removal of charged species in the upper polar atmosphere is described. A large-scale efficient charging system could be beneficial to the removal of CFCs in the troposphere through dissociative attachment.

I. BASIC CONSIDERATIONS

The first photoelectric measurement of the amount of ozone in the atmosphere was made by Dobson and Harrison.[5] The nonuniform distribution of the ozone in the atmosphere is maximum in the stratosphere (at an altitude of approximately 25 km, depending on the latitude). The first theoretical explanation for the ozone layer was given by Chapman[6] who proposed a pure-oxygen photochemical model for the generation and destruction of ozone. His reaction scheme is described by the following set of equations:

$$O_2 + h\nu \rightarrow O + O \tag{1}$$
$$O + O_2 \rightarrow O_3 + M \tag{2}$$
$$O_3 + h\nu \rightarrow O + O_2 \tag{3}$$
$$O + O_3 \rightarrow 2O_2 \tag{4}$$
$$O + O + M \rightarrow O_2 \tag{5}$$

Equation 2 describes the formation of an ozone molecule in a three-body collision of an oxygen atom with an oxygen molecule and a third collision partner M.   Equation 3 is the photodissociation of ozone. Equation 4 is the loss of one oxygen atom and one ozone molecule in a collision leading to the formation of two oxygen molecules. The last reaction, Eq. 5 is the loss of two oxygen atoms in a three-body collision resulting in the formation of one oxygen molecule.

Although Chapman's model can explain the formation of the ozone layer, the pure-oxygen model leads to calculated values for the ozone abundance in the atmosphere which are higher than experimental results.[7] It was concluded that an additional loss mechanism is affecting the ozone concentration which is not included in the pure oxygen-model. Researchers have been considering catalytic reactions of trace constituents of the atmosphere with ozone as an additional loss mechanism. The simultaneous observation[5] of ozone depletion and increased concentration of chlorine oxide in the stratosphere suggested that catalytic reactions of chlorine atoms with ozone molecules contribute to the loss of ozone. The following reaction chain has been proposed as a dominant catalytic cycle:

$$Cl + O_3 \rightarrow ClO + O_2 \tag{6}$$
$$Cl + O \rightarrow Cl + O_2 \tag{7}$$

The net effect of reactions (6) to (7) is the conversion of ozone molecules into oxygen molecules.

Molina and Rowland[8] showed that atomic chlorine in the contemporary atmosphere primarily originates from the photolysis of two chlorofluorocarbons Cl_3CF (CFC–11) and Cl_2CF_2 (CFC–12). Both substances are produced on an industrial basis and have no natural origin. They are very inert and volatile. Their estimated lifetime in the atmosphere ranges from 40 to 150 years. They are released at the surface of the earth and transported by convection into the stratosphere, where the exposure to UV radiation leads to their photodissociation into atomic chlorine and the radicals Cl_2CF and $ClCF_2$, respectively. The atomic chlorine can then participate in the catalytic reactions (6) to (7) which destroy ozone.

Thermodynamic Considerations

We would like to demonstrate with thermodynamic arguments that the destruction of ozone by the reaction of free chlorine atoms with ozone molecules can significantly be reduced if the chlorine atoms are converted in negative chlorine ions Cl⁻. The change in Gibbs' free energy ΔG for the reaction of neutral chlorine atoms with ozone molecules

$$Cl + O_3 \rightarrow ClO + O_2 \qquad \Delta G = -168 \; kJ/mol \qquad (8)$$

is -168 kJ/mol, while the change in free energy for the reaction of negative ions with ozone is only -13kJ/mol.

$$Cl^- + O_3 \rightarrow ClO^- + O_2 \qquad \Delta G = -13 \; kJ/mol \qquad (9)$$

The small change in the free energy of the reaction of Cl⁻ with ozone is effected by the different electron affinities of Cl atoms and ClO molecules, which are 3.7eV and 2.0eV, respectively. The small change in the free energy of the Cl⁻ reaction with ozone leads to a reaction rate of less than 10^{-12} cm³/sec which renders this reaction ineffective for the destruction of ozone in the atmosphere.

In order to account for all the important species, which participate in the atmospheric reactions with ozone, we are using computer modeling; however, the computer results must be checked against laboratory experiments. Computer modeling is better at explaining observations than making predictions; e.g. the ozone hole was not predicted.

II. COMPUTER MODELING AND LABORATORY EXPERIMENTS

A computer model of atmospheric transport and kinetic chemistry, developed by J. Zinn and C. D. Sutherland at LANL,[9] was modified to include all known ion reactions related to ozone. The original code was used to simulate the atmosphere over a wide range of temperatures. It can model the atmosphere with a dimensionality ranging from 0 to 2, where a zero-dimensional atmosphere has no transport. The model predicts the densities of the chemical species in the atmosphere as a function of time, given the electron and neutral temperatures, the initial densities of the species, and the rate at which the species enter the atmosphere.

The present input consists of 62 species, including molecules and ions, and will expand as additional chemical processes are considered in the model. The computer program searches its library file for those reactions whose products and reactants are subsets of the input species. A smaller library file, formed by the computer program, contains only the reactions which meet this criterion. For the present input file, the smaller library file consists of 1104 reactions. The large library file contains reactions of the following types:

- ionizations and recombinations
- attachments and detachments including photodetachments.
- charge transfer
- associations and dissociations
- excitations and de-excitations
- atomic interchange and rearrangements

These reactions are taken from D. Smith and M. J. Church,[10] and a variety of other sources. The model will soon include the kinetics of negative cluster ions, as in G. Brasseur and A. Chatel.[11] When reverse reaction rates are unknown, they are calculated by detailed balancing.

All of the reactions included in the smaller library file combined with sources and sinks constitute for each species a nonlinear first-order differential equation of the form:

$$\frac{dN_i}{dt} = \sum_{j,k} K_{i,j,k} N_j N_k + Q_i(t) \tag{10}$$

where N_i and N_j are the number densities of species i, j, and k respectively, $Q_i(t)$ is the contribution of external sources, $K_{i,j,k}$ are rate coefficients, and the sum is over all reactions involving species i, including two-body, three-body, and photo-reactions. Using discrete time increments and matrix inversion, the program solves the many simultaneous equations of the above form as a function of time, thus yielding the time evolution of the density of each species. The time increment can be adjusted to accommodate processes with short and long time scales (from microseconds to millions of seconds).

Equilibrium is obtained by letting the system evolve for more than 10^8 seconds, until the rate of change of N_i is less than 0.01 % per 10^6 seconds, for all i. The effect of a perturbation of the atmosphere, such as an influx of chlorine or electrons, can be determined by perturbing the equilibrium, and letting the system evolve to a new equilibrium. The equilibrium can be perturbed through either a pulsed or continuous injection of various species, or a change in temperature.

In addition to listing the density as a function of time, the output gives the 12 most important reactions for each of the species, as a function of time. These are the reactions occurring at the highest rate per unit volume and unit time. The output is downloaded from the Cray computer to a PC for plotting.

A 40 km baseline atmosphere is established in the model starting with the actual observed concentrations for the major species, which remain fairly stable as the model is integrated for 10^5 seconds (over 1 day). The effects of increased active chlorine are best simulated by increasing the concentration of ClO, rather than increasing the Cl density. This method is justified because the concentration of the former species is more than 100 times larger than that of the latter at 40 km, and the rapid interchange between Cl and ClO assures that this ratio will be preserved for moderate increases in the concentration of either species. Any modest increase in Cl will be immediately used up by the much larger ClO concentration, while a significant percent change in ClO will automatically be applied to Cl.

The increased catalytic destruction caused by additional active chlorine is demonstrated by re-running the simulation with an initial increase in the concentration of ClO by 23%. After 10^5 seconds, both Cl and ClO concentrations are still increased by 23%, and the increased catalytic destruction results in a decrease of 10% in the ozone concentration. The time dependence of this behavior is shown in Fig. 1. It should be noted that the reason for the small, relative decrease in ozone is that NO is also a catalyst of ozone destruction, which is equal in importance to active chlorine at 40 km.

The beneficial effect of negative ion conversion of active chlorine is then demonstrated by re-running the 23% increased ClO simulation with an initial injection of 10^7 electrons/cc or completely negating the effect of the 23% increase in ClO. The time evolution of Cl and ozone is shown in Fig. 2. Additional computer modeling[12,13] has shown the efficiency of electron attachment to neutral chlorine atoms.

Laboratory Experiments

In order to conduct controlled studies of catalytic destructive processes and mitigation measures in the upper-atmosphere, we have built a laboratory chamber[14] of sufficient dimensions in which ozone and reactive species have sufficient lifetimes to react repetitively. The neutral pressures and the UV spectrum are chosen to simulate the stratospheric conditions and at the same time satisfies detectability. The simplified Chapman model is examined first with a controlled injection of minority species. The catalytic destruction by Cl dissociated from CFC is verified experimentally. When charges are injected into the medium and Cl⁻ ions are formed the recovery of the ozone proceeds as chlorine radical is disabled and collected. In subsequent experiments additional species, one at a time, will be included to ascertain their contributions to the ozone depletion or recovery.

The Reaction Chamber

A schematic drawing of the reaction chamber is shown in Fig. 3. The chamber has an inner diameter of 1.40 m, and its overall length including the two extensions is 4.03 m. The chamber can be evacuated by a turbomolecular and a rotary vane pump to a pressure of 10^{-5} Torr. For operations under static conditions, the pumps can be valved off with a gate valve. The main vessel is equipped with several types of vacuum gauges to permit pressure measurements over the range from hundreds of Torr to 10^{-5} Torr. An ozone generator is connected to the chamber, which can produce a mixture of 2.5% ozone and 97.5% oxygen at chamber pressures ranging from a few mTorr to hundreds of Torr. This external injection of ozone is mainly for calibration purpose. Internal solar simulators are used to generate ozone by their radiation in the UV range.

The ozone concentration n in the chamber is determined from absorption measurements of the 253.7 nm line of mercury, which is almost at the center of the Hartley band of ozone. The ozone density, n, can be obtained from the equation:

$$I = I_0 \exp (-l\sigma n) \qquad (11)$$

where l is the optical path length of the analyzing beam in the chamber, σ is the absorption cross section, and I and I_0 are the transmitted and the unattenuated intensity, respectively. Absorption cross sections for this line have been measured by several authors.[15,16] For absorption measurements, a light beam emitted from a low pressure mercury lamp is passed through the reaction chamber, and the transmitted intensity of the 253.7 nm component is detected with a monochromator and a photomultiplier tube. The light beam is chopped at a frequency of 215 Hz, and before the beam enters the chamber, a portion of its intensity is deflected by a beam splitter into a control monochromator, which monitors the lamp intensity at the 253.7 nm line. A lock-in amplifier locked to the chopping frequency of the light beam is used to determine the ratio of the photomultiplier signal to the output voltage of the control monochromator. Measuring this ratio eliminates fluctuations in the absorption measurements caused by the slight variations in the intensity of the light source. For studies of photochemical processes, the reaction chamber is provided with two 140 W mercury lamps (Light Source Inc. Model GSL 1524 T5 VH/HO) which are the solar simulators. The lamps are mounted inside the chamber and have significant intensity at the 185 nm line. The lamps are operated with 60 Hz line voltage, and their intensity is therefore modulated with 120 Hz. The high selectivity of the lock-in amplifier prevents any interference of the mercury lamps with the intensity measurements of the analyzing beam.

Several portholes in the vacuum envelope of the chamber permit access of optical spectrometers to the chamber for spectroscopic investigations in the visible UV, and vacuum-UV region of the spectrum. A mass spectrometer connected to the chamber allows independent determinations of the reaction products.

Experimental Procedures and Results

The photochemical generation of ozone was measured at oxygen pressures from 0.5 to 10.1 Torr when the chamber was valved off from the pumps. Since electric breakdown voltages are very low in this pressure regime, the UV light sources were started when the reaction chamber was still evacuated in order to avoid the high start-up voltage of the mercury lamps at almost critical pressures. This procedure was followed during the first measurements. During later experiments, we were able to start the light sources when the chamber was already filled with 10 Torr of oxygen. When oxygen was leaked into the chamber in the presence of UV light, the transmitted intensity of the 253.7 nm line in the analyzing beam decreased exponentially with the ozone density n according to Eq. (11). A plot of the transmitted intensity of this line as a function of time yields the time evolution of the ozone density as shown in Fig. 4. At point A, the flow of oxygen into the chamber was started, and the sharp drop of the transmitted intensity indicated a sharp rise in the formation of ozone according to the reactions in Eqs. (1) and (2). At point B, a pressure of 10.1 Torr was reached, and the flow of oxygen was stopped. The ozone density continued to increase until an equilibrium was reached at point C.

From measurements of the time rate of change of the ozone density and its equilibrium values we can deduce in a self-consistent manner the rate of generation and the rate of destruction in response to various additions of trace constituents or change in the environment.

In addition to the ozone losses described by Eqs. (3), (4), and (5), ozone was lost in the reaction chamber during collisions of ozone molecules with the chamber wall. This lost rate was measured by turning the solar simulator off. A typical loss of ozone density due to the wall effect was 6.5 x 10^9 (cm^{-3} sec^{-1}), which is much less than the typical reaction rates of 3 x 10^{11} (cm^{-3} sec^{-1}) among species inside the chamber, and therefore permits repetitive reactions to occur. The equilibrium ozone concentration at point C in Fig. 4 was 4.6 x 10^{14} cm^{-3} i.e. approximately 0.1% of the O_2 molecules were converted into ozone molecules.

At point C of Fig. 2 approximately 3 mTorr of chlorofluorocarbon (also known as trichlorofluoromethane Cl_3CF) vapor were leaked into the chamber, and we observed a rapid decrease of the ozone density. This loss of ozone was observed only when Cl_3CF vapor was irradiated by UV light from the solar simulator. We deduced that the ozone destruction was caused by the reaction of ozone with products of photodissociated Cl_3CF molecules. As it had been proposed for the generation of atomic chlorine in the stratosphere, the photolysis of Cl_3CF by UV radiation resulted in atomic chlorine and the radical Cl_2CF. The ozone loss was caused by reactions of atomic chlorine with ozone according to Eq. (6).

Since atomic chlorine is restored in the subsequent reactions of the product molecule ClO, Eq. (7), the relatively small addition of Cl_3CF to the gases in the reaction chamber significantly reduced the ozone concentration. The catalytic destruction of ozone by atomic chlorine led to the establishment of a new equilibrium of ozone generation and destruction at point D. The ozone concentration at point D was reduced to 5.7 x 10^{13} cm^{-3}, which is only 12% of the concentration prior to the addition of Cl_3CF. A plot of ozone concentration n as a function of the density of Cl_3CF is shown in Fig. 5. The ozone concentrations were measured at the equilibria between ozone generation and destruction, which we observed for increasing amounts of Cl_3CF vapor in the reaction chamber. The partial pressure of Cl_3CF was measured with a differential capacitance manometer (MKS Instruments Inc., Model 120AD), which permitted measurements of pressure changes of the order of 3 x 10^{-4} Torr when the pressure in the chamber was 10.1 Torr. At the lowest measurable change in partial pressure, 0.3 mTorr, the curve in Fig. 5 exhibits a very sharp drop of the ozone concentration n, followed by a slower decrease at higher CFC densities n_c. The data points could be fitted by a function of the form

$$n = 1/(An_c + B) \tag{12}$$

where the coefficients A and B were obtained from at least a square foot and the constraint that the function passes through the data point for $n_c = 0$. This constraint was chosen because the data point for $n_c = 0$ was obtained with much higher accuracy than all other points. The following values were obtained for A and B: $A = 1.833$ (10^{-28} cm^6), $B = 0.161$ (10^{-14} cm^3). The solid line in Fig. 5 was obtained from Eq. (12) with the above-given values for the coefficients.

We can explain the functional dependence of the equilibrium ozone concentration n on the number density n_c of CFC molecules as follows:

$$dn/dt = k_z Jn_{02}^2 - v_w n - k_c n n_{Cl} \tag{13}$$

where $k_z n_{02}^2$ is the quantum yield for the generation of ozone, v_w describes the wall losses, k_c is the rate constant for ozone destruction by chlorine atoms, and n_{02} and n_{Cl} are the number densities of molecular oxygen and atomic chlorine, respectively; J is the photon flux. Under equilibrium conditions

$$dn/dt = 0 \tag{14}$$

we find the relation

$$n = \left(\frac{k_c n_{Cl}/n_c}{k_z Jn_{02}^2} n_c + \frac{v_w}{k_z Jn_{02}^2} \right)^{-1} \tag{15}$$

which has the same form as Eq. (12). Values for $k_z Jn_{02}^2$ were obtained from measurements of the slope dn/dt at the beginning of the ozone generation when ozone losses are still negligible, and values for n_{Cl}/n_c were obtained from measurements of the slope of n after injection of trichlorofluorocarbon vapor into the reaction chamber. For k_c we used a published rate constant, the value of which at room temperature is 1.189×10^{-11} cm^{-3} sec-1 and we obtained v_w from the ozone decay rate after shutting off the UV light sources. With these parameters, we obtained for the coefficients A and B the following values:
A = 1.08 (10^{-28} cm^6), B = 0.072 (10^{-14} cm^3). The calculated values of A and B differ from the measured values by a factor of approximately 2. The theoretical estimate according to Eq. (15) is shown in Fig. 5 as a dotted curve.

From our measurements of n_{Cl}/n_c we can deduce the multiplicative destructive factor of ozone molecules by Cl radical. Using the average ratio of $n/n_c \approx 1$ we find that one chlorine atom destroys up to 4×10^5 ozone molecules.

Simultaneous Injection of CFC and Charges

To prove the concept that charge injection is indeed beneficial, we have installed an RF plasma source which injects charges into the large reaction chamber simultaneously.[17] This source yields an electron flux of 0.14 mA over a cross section of 26 cm^2. For an initial pressure of O_2 of 5 Torr, we achieve a density of ozone of 1.2×10^{14} cm, to which 0.1 mTorr of CFC is injected. There is an immediate drop (within 100 secs) of ozone density down to 25% of its initial value. If the same experiment is repeated with the injection of charges as mentioned above, the ozone density recovers its full initial value, in approximately 90 min. By repeating this controlled experiment a number of times we have verified that the charge injection accounts for the recovery of the ozone level in the chamber. Computer modeling of the above conditions shows the successive transfer of the charges from O^- to Cl^- in approximately the same period.

III. METHODS OF ATMOSPHERIC CHARGING

The concept of atmospheric charging can begin from the fact that the earth is negatively charged and possesses what is known as a "fair weather" electric field which varies diurnally between 100-130 V/m at the earth's surface. This field intensity means that the earth carries a total electric charge of -500,000 C. The ionosphere is oppositely charged, with the result that the earth-ionosphere combination can be thought of as a huge spherical capacitor, leaking electric charges through the atmosphere. The potential difference between the earth and the ionosphere is 300,000 V, and the total leakage current is 1500-2000 A, giving a total atmospheric resistance of only 200 - 150 Ω. To maintain the "fair weather" field, the ionosphere has to be fed by other sources of current. These sources turn out to be the thunderstorms that occur daily all over the earth, primarily over the Amazon and African land masses.

Within thunderclouds, charges are separated by the formation of water droplets, ice crystals, and gravity together with convection, acting on these particles. The top of the thundercloud becomes positively charged, the middle negatively charged, and the bottom of the thundercloud slightly positively charged. These charged regions build up until there is a discharge to the ground, leaving the top of the cloud, positively charged. This excess charge redistributes itself by repulsion throughout the ionosphere, contributing to the maintenance of the "fairweather" electric field. Other thunderclouds make similar contributions. It is estimated that over the earth's surface there are 1,800 thunderstorms in process at any one time.

Large scale charge separations are, thus, possible in the atmosphere; however, they require the formation of particulates (water droplets, ice crystals and larger ice particles referred to as "graupel"), and the action of gravity on the heavier particles to cause them to fall and separate away from the lighter, oppositely charged particles.[18] The properties of water also play a prominent role in this process through its electrical polarizability and the generation of electrical potential during changes between the solid and liquid phases.

In the following we will discuss two schemes of artificially electrifying the atmosphere, by (a) the simultaneous release of positive and negative charges, and by (b) the injection of electrons through a current channel.

(a).   Simultaneous release of positive and negative charges.

Taking a clue from thundercloud phenomena, a scheme has been devised by Lawrence Livermore Laboratory (LLNL) and UCLA scientists, C.D. Hendricks and R. F. Wuerker, for electrifying the atmosphere with either an airplane, a rocket or a balloon. The scheme is shown schematically in Fig. 6. It is based on having a long wire or antenna biased at a sufficiently high negative voltage to emit electrons by field emission. In addition, the plane carries a water supply which releases water droplets with the opposite polarity. Charging of the droplets occurs by sharp hollow needles at the bottomside of the aircraft or balloon. The release of charges of both polarities ensures that the plane remains uncharged and a steady deposition of negative charges in the atmosphere is possible. The water droplets are large enough to fall under gravity, causing a separation between opposite charges. Separation will occur when the emitted charges or ions around the antenna form small hydrated clusters or submicroscopic particles of low mobility. The very processes that are set up by nature in thunderclouds are being attempted in the aforementioned scheme.

The atmospheric electrification scheme shown in Fig. 6 works in either polarity, the antenna can be biased positively and the neutralizing water-emitting needles biased negatively, creating a positively charged cloud, or vice versa.

The neutralizing droplet size is determined by the size of the internal diameter of the needles, being approximately twice the needle diameter.

It can be shown that the maximum charge q_{max} on each individual water droplet is

$$q_{max} = 8\pi\sqrt{\varepsilon_o T r^3} \qquad (16)$$

where $\varepsilon_0 = 8.85 \times 10^{-12}$ F/m (permittivity of free space), T = surface tension of water (0.075 N/m), and r = droplet radius in meters. For example, a droplet of 10 μm radius could be charged to 4 x 10^6 elementary charges or carry away 0.6×10^{-12} C per particle. This equation is an upper bound since it expresses the fact that within the droplet the forces of electrical repulsion are just balanced by the forces of surface tension. Any more charge on the droplet would cause it to divide or emit the excess charge as a jet.

The charge to mass ratio of such a particle is easily computed and is the more significant number;

$$\left[\frac{q}{m}\right]_{max} = \frac{6}{\rho}\left[\frac{2\varepsilon_o T}{r^3}\right]^{1/2} \qquad (17)$$

where ρ is the density of the droplet which is 1000 kg/m³ for water. Thus, for our aforementioned 10 μ diameter water droplet ejected from the bottom of the airplane, the maximum charge to mass ratio would be 0.15 C/kg. This value can also be interpreted as 0.15 C of charge ejected by the antenna for each kg of water dropped away by the neutralizing needle.

Maximum charge values for water droplets of different sizes are included in Table I, which also includes values for the electric field strengths at the surface of the particle. The limiting charge values and charge to mass ratios apply to vacuum conditions where there are no collisions with neutral particles which might carry charges away. The limiting electric field for conductors in air at atmospheric pressure is 30 KV/cm.

TABLE I

Water Droplet Electrification

Drop Radius (microns)	Max Charge (10^{-12} C)	Max Charge/Mass Ratio (C/kg)	Surface Field (KV/cm)
10	0.6	0.15	600
20	1.8	0.053	150
30	3.2	0.028	67
40	5.0	0.019	38
50	7.0	0.013	24
60	9.2	0.010	17
80	14.1	0.007	12
100	20.0	0.005	6
200	60.0	0.0017	1.5

Particles will fall with a limiting velocity (v_{term}) determined by the viscosity of air and the cross section of the particle as specified by Stokes law;

$$v_{term} = 2\rho g \ r^2/9\eta \qquad (18)$$

where η is the viscosity of air (1.83×10^{-5} kg/m sec at 23°C)

Stoke's laws gives a terminal velocity of 1 cm/sec for water droplet of 10 micron radius, as shown in Table II. Since viscosity is approximately independent of pressure, the terminal velocity applies over a wide range of altitudes.

TABLE II

Terminal Velocities of Water Droplets in air at 23°C

Radius (microns)	Terminal Velocity (cm/second)
10	1
20	4
30	9
40	16
50	25
60	36
80	64
100	100

The neutralizing particles, being rather highly charged, can be expected to grow larger in falling through layers of saturated water vapor, thus increasing their downward velocity. Such processes occur in thunderclouds.

Generation of Electrons by Photoelectric Emission

The most efficient method of injecting low energy electrons into the atmosphere is to use solar radiation whose flux density is approximately 1kW/m². The photoelectric emission of electrons from a metal surface such as zinc has been found to be of the order of 1μA/W which is equivalent to the injection flux of 10^{13} electrons/sec. Assuming a photocathode area of 10m² emitting into a volume of $10^3 m^3$, one finds, for the rate at which the electron density changes, the value of $10^5/cm^3$sec. This value is much greater than the rate of density increases which have been considered in the computer modeling of the previous sections.

The rate of electron injection is determined by the rate at which positively charged droplets can be released by the space vehicle. Since each 10μ droplet can carry a maximum quantity of 4×10^6 elementary charges, a release of 10μ particles at the rate of 2.5×10^7/sec is required.

(b). Injection of electrons through a current channel.

Instead of injecting positive and negative charges simultaneously we can circulate the electrons through a diode, which includes a cathode of low work function. Electrodes of low work functions(

2.2-2.6 ev) can be made by coating the surface with sodium or barium, protected from the oxygen atmosphere with a gold coating.[19] In the laboratory under simulated conditions solar UV radiation has been observed to produce an emission current density of $1\mu A/cm^2$ or 10^{13} electrons/sec-cm^2. Even in the absence of solar radiation it is energetically favorable for halogens such as chlorine atoms (electron affinity of 3.7 eV) to acquire electrons from an electrode of lower work functions upon contact. These low energy electrons will attach themselves to the majority species of oxygen atoms and molecules first. Successive collisions between oxygen ions and chlorine atoms transfer electrons from O^- to form Cl^- because of the large difference in their electron affinities.   Once formed, the negative Cl ions drift to the positively biased anode and are collected to form metallic chlorides.

The electron given up by Cl^- at the anode surface is circulated through the electrical connection to the cathode and the process can be sustained in a steady-state manner without charge buildup. A negative charge density in the form of electrons and negative ions is maintained in the current channel.

Consider a diode module with a separation of 1 meter between electrodes. A voltage of 1 Kv is imposed on this diode configuration by a power supply derived from solar energy flux (1 kW/m^2). The dc electric field inside this electrode is given by

$$E_o = V_o /L = 10V/cm \tag{19}$$

Assuming ion neutral collision cross section of 10^{-16} cm^2, the mean free path λ for an atmosphere of 1 Torr is $\lambda = (n\sigma)^{-1} = 1$ cm and the drift velocity v_d

$$mv_d^2 = Ee_o\lambda \text{ or } v_d = 5 \times 10^5 \text{ cm/s} \tag{20}$$

for each 1 m^2 anode area.

For a chlorine density of $10^6/cm^3$ the collected Cl^- flux = 5×10^{15} sec^{-1} for each square meter of collecting area. The flux collected by a large panel, each having an area of 500 m x 500 m, amounts to 1.5×10^{21} sec^{-1}.

The above calculation is for a pressure of 1 Torr. At the lower height the mean free path is shorter with increasing atmospheric pressure. We would need to increase the voltage and the area of collection. The drift velocity of ions is proportional to the square root of the mean free path and the applied electric field. In the following section we will consider a method of charging and collecting particulates in the Polar Stratospheric clouds (PSC). These clouds are believed to be depositories of pollutants and provide surfaces for heterogeneous chemical reactions.

Global Content of Active Cl between 30 - 45 km

We wish to consider the global content of active Cl and then the time required to collect a significant portion of Cl or their reservoirs. Taking a mixing ratio of Cl as 2.4×10^{-10} and a air column density of $2.1 \times 10^{23} cm^2$, the total active Cl column density is $5.2 \times 10^{13} cm^2$ for a global surface area of $5.15 \times 10^{18} cm^2$. The total number of active Cl in the shell of 15 km height is 2.7×10^{32} with a total weight of 1.5×10^{10}g or 1.5×10^4 ton.

Collection of Particulates in the PSC

There is strong evidence that PSC is a source of chlorine atoms[20] in Antarctica. The method outlined above is also adaptable to charging and collecting the particulates within these clouds which have been found to be at or above an altitude of approximately 12 km. The particulates are typically of 3 micrometer in diameter and contains HCl of 1% by weight. The density of these particulates in PSC has been estimated to be of the order of 10 cm^{-3}. If the ambient gas moves at a velocity of 30 m/sec with respect to the platform, the flux of particulates collected by an area of 1 m^2 is given by:

$$Ip = nva = 10 \times 3 \times 10^3 \times 10^4 = 3 \times 10^8 sec^{-1} \tag{21}$$

Since each particulate is the reservoir of approximately 10^9 chlorine atoms, the flux of chlorine collected is

$$I_{Cl} = 3 \times 10^{17} sec^{-1} \tag{22}$$

If we take a collection area of a panel, 500 m x 500 m or 2.5 x 10^5 m^2, the number of particles collected per panel = 2.5 x 10^5 x 3 x 10^{17} = 7.5 x $10^{22} sec^{-1}$. If each polar region contains 10% of the total chlorine reservoir or 2.7 x 10^{31}, the time required is 3 x $10^8 sec$ or approximately 30 years with one platform. With 10 platforms in each polar region we can achieve this within 3 years. An obvious advantage of collecting PSC particulates is that their locations can be determined by laser scattering. Since these clouds can only form at locations where the temperature is below a certain threshold (190° K) their approximate locations can be predicted several days ahead by using computer modeling to project the temperature profile in the polar region.[21]

The general collection strategy is to recognize that the polar regions are the depositories of chlorine compounds which are transported up there by large scale convections from mid-latitudes. Furthermore within the polar region only the location of PSC clouds need be treated, therefore further reducing the remediation region. The remediation region is estimated to be limited to a small manageable portion (< 0.5%) of the global surface. The number of space platforms is correspondingly reduced.

IV. MIDDLE ATMOSPHERIC PLATFORM (MAP)

Two conceptual designs of an unmanned Middle Atmospheric Platform (MAP) are depicted in Figs. 7a, 7b. As shown in Fig. 7a a single airship has a panel suspended from it of dimensions 0.5 km x 0.5 km, consisting of a number of cylinders with each a central electrode. The total collection area, considering collection on both sides of each panel, is 5×10^5 m^2. The central electrode is biased negatively at 3 kV and current flows between it and the outer cylinder. Using light weight fibre for the construction of the panel a total weight of 2 metric tons for the load is estimated. Figure 7b shows a larger platform design with multiple floatation units and a larger solar collection area. At a speed of 30 m/sec or approximately 100 km/hr with respect to the ambient atmosphere, the volume of collection per day is approximately 1000 km^3. The shape of the platform has been optimized to reduce the drag. Engineering designs of such platforms has been performed by *Deutch Aerospace*[22], Lockheed in HISPOT[23] and by MITI of Japan in HALROF.[24]

Unlike balloon platforms, each MAP platform is navigable by a propulsion system appropriate to the height of operation. At stratospheric heights the thinner atmosphere requires ion engines mounted at the rear of the platform. This special ion engine is solar powered and uses the ambient atmosphere as the source of gas, thus saving the payload for fuels. It can be shown that the momentum imparted by ions to the platform can balance the drag of the space platform in prevailing winds, taking the aerodynamic shape of the hull into account. At altitudes of 12 km and lower, conventional engines will be used. Extra helium tank and storage battery round out the required payload for long duration operation.

Breakup of CFCs Through Electron Attachment in the Troposphere

CFCs can break up readily upon contact with low energy electrons.[1,25] There is an advantage of dissociating CFC molecules in the troposphere on their way up to the stratosphere, if there is an economical method of generating charges on a large scale. In the troposphere solar radiation of 300 nm and longer can release electrons from the surfaces of emitters of low work function (< 4 eV). Electrodes of protective coating against oxidation can be made and are being fabricated in our laboratory and tested under simulated tropospheric conditions for their endurance, in the presence of oxygen and uv radiation of relevant wavelengths. These electrons will be attached to the majority species first such as O_2, which then collide with CFC molecules, resulting in their dissociation. These negative charges are maintained in a conducting channel as described in Section III. Since the current is required to be continuous and is carried by negative charges, the recombination is not a limiting factor.

A proposal using convergent focussed microwaves beams to generate electrons through ionization has been made.[25] A laboratory experiment[25] has demonstrated that CFC can be reduced in such a discharge. However the energy required for the dissociation of CFCs amounts to 1 Kev/molecule and will be rather prohibitive from a cost standpoint. Furthermore the recombination in a high density discharge will affect the overall efficiency. In comparison the use of solar energy and a large emission surface on a MAP platform combined with the concept of a current channel consisting of one sign of charges might make a large-scale breakup scheme realizable in the troposphere.

V. SELECTIVE REMOVAL OF IONS BY ION CYCLOTRON ACCELERATION IN THE POLAR REGION

Although in previous sections we have described how to effect in-situ collection of pollutants in the stratosphere, there is another approach to expelling pollutants out of the upper atmosphere using electromagnetic waves in plasmas. Waves influence particles over a much larger region of space than physical platforms.

Laboratory experiments[26] have shown that significant acceleration of a minority ion species at its cyclotron resonance is possible because the induced current carried by the minority species can be canceled by that of the majority species. Based on the above principle and the morphology of the magnetic field of the polar region, a method of removal of pollutants, which are the minority species in the upper atmosphere, is described.[27,28] In the polar region negatively charged species are moved upward by the downward-pointing fair-weather electric fields, which can be artificially enhanced by ionospheric modification[29] or by atmospheric convection arising from the temperature profile between the stratosphere and the mesosphere. Once they reach the height (above 100 km) where the cyclotron frequency is comparable to the collisional frequency, they can be resonantly accelerated by EM fields, oscillating at their ion cyclotron frequency.

These extremely low frequency EM fields (called ELF waves), in the ion cyclotron frequency range(20-50 Hz), corresponding to major and important minority species at the earth's magnetic field of 0.5 gauss, have been successfully excited over a large horizontal extent (30-100km) in the upper atmosphere.[30] This was accomplished by beaming high power HF electromagnetic waves, which are modulated at ion cyclotron frequencies, towards the auroral electrojet region which extends from 75 km to 120 km (known as the D and E regions).[31] These EM fields heat up electrons and modify the conductivity of the D and E region at their modulation frequency. The background DC electric field in the ionosphere then drives an oscillating current, which is the source of the EM ion cyclotron wave, propagating up and down the earth's magnetic field lines.

We can write the refractive index K of the ion cyclotron waves including the minority species, carrying charges of either sign for both the right-hand and left-hand propagating waves including collisions as follows:[27,28,32]

Using cold plasma theory the reflective index K is:

$$K^2 = R \cong C \left[+1 - \sum_i \frac{a^+}{1 + \beta_i^+ + i\gamma_i^+} - \sum_j \frac{a^-}{1 - \beta_j^- + i\gamma_j^-} \right] \text{ right-hand propagation} \qquad (23)$$

$$K^2 = L \cong C \left[-1 + \sum_i \frac{a^+}{1 - \beta_i^+ - i\gamma_i^+} - \sum_j \frac{a^-}{1 - \beta_j^- + i\gamma_j^-} \right] \text{ left-hand propagation} \qquad (24)$$

$\sum_i$ summation over positive ions species, $\sum_j$ summation over negative ion species

$$\beta^+ = \frac{\omega}{\Omega^+} \quad , \quad \gamma_i^+ = \frac{v_i^+}{\Omega^+} \quad , \quad a^+ = \frac{n^+}{n_e} \qquad (25)$$

$$\beta^- = \frac{\omega}{\Omega^-} \quad , \quad \gamma_j^- = \frac{v_j^-}{\Omega^-} \quad , \quad a^- = \frac{n^-}{n_e} \qquad (26)$$

$$C = \frac{\omega_{pe}^2}{\omega\Omega_e} \qquad (27)$$

For negative ions the right-hand wave always propagates and accessibility is assured. For positive ions the left-hand wave propagates near each of ion cyclotron resonance. The amplitude of the excited ELF wave is inversely proportional to the square root of the refractive index, which is a minimum as a result of the cancellation between the induced currents of the majority and minority species.

Using observed ELF magnetic fields of 1pT by ground-based detectors and associated electric field strengths of 300 microvolt/m, we have deduced a wave field of 10 pT and 3 mV/m at the 100 km height which is closer to the excitation region. The excited EM ion cyclotron wave propagates mainly along the magnetic field with a transverse dimension of 100 km, as determined by the beam width of the primary high-frequency EM field. A negative Cl ion can be resonantly accelerated to the escape energy of 23 eV in approximately 10 sec and will move upward along the magnetic field by the $\mu\nabla B$ force. The earth's magnetic field lines are distorted by the solar winds and the nearly vertical field lines become open field lines and extend well into outer space, even beyond the magnetosphere. Particles accelerated upward along these vertical field lines will not return into the atmosphere.

This concept of ion acceleration is supported by the observation of ion conics over the polar regions. Satellite detectors have recorded upward moving oxygen ions of energy in the keV range, which is believed to be the result of transverse acceleration by naturally generated plasma wave instabilities near the fundamental or harmonics of the ion cyclotron frequency or the lower hybrid frequency.[33] The same free energy which excites these instabilities can be channelled into a coherent ELF wave. Laboratory experiments have definitively shown that a small artificially injected coherent signal, ten times the thermal fluctuation level, is preferentially amplified by free energy sources to become a large-amplitude single mode in place of an otherwise broad-band low-amplitude unstable spectrum. This is an universal phenomenon and should be applicable to the E region of the ionosphere and might lower the overall requirement of the ground-based transmitting facility.

This ion cyclotron acceleration method does require the transport of negative ions or their hydrates to a height where they remain negatively charged and become collisionless. A detailed consideration of the diffusion and convection of these particles to greater heights as a result of the temperature and profile between the stratosphere and the mesosphere, will be given in a separate paper.

Assuming a chlorine density of $10^4 cm^{-3}$ in this region and a vertical resonance region of 10km, we have found that a total of 10^{24} particles can be accelerated in 10 sec. The time required to clear the entire polar region consisting of 5.4×10^{30} chlorine particles is $5.4 \times 10^{30}/10^{23} = 5.5 \times 10^7$ sec or approximately 1.7 year. This would require ELF power of 400KW, which must be supplied by at least 40 radiating facilities with a conversion efficiency of 1% from HF power to ELF frequencies. While this efficiency has not been achieved yet, research into the channelling of the free energy sources, such as electron currents along the polar magnetic field lines, into wave generation might make this possible.

CONCLUSION

We have shown the beneficial effects of charging the atmosphere by laboratory and computer modeling studies. Ions, molecules and clusters, which are charged, respond to external electromagnetic fields and are therefore subject to remote control. A number of laboratory experiments still need to be conducted to find the most efficient method of charging the atmosphere. Here we can learn much from the existence of large-scale charged clouds in nature and their slow recombination rates. A rigorous effort in plasma physics and space physics applied to the polar atmosphere will bring progress to our research on the mitigation of the worsening depletion of the stratospheric ozone layer.

ACKNOWLEDGMENTS

I wish to acknowledge the contribution of Dr. Ralph Wuerker in the design of the large platform with the collector cage. Dr. K. N. Leung of Lawrence Berkeley Laboratory has given the suggestion of using barium coated surface to reduce the work function. Discussion with Drs. R. Suchannek, D. Sensharma, G. Rosenthal, D. Ho are acknowledged. This work is supported by the National Science Foundation, the UCLA Department of Physics and the Ozone Society.

REFERENCES

1. A. Y. Wong, J. Steinhauer, R. Close, T. Fukuchi, and G. M. Milikh, Comments Plasma Phys. Controlled Fusion, 12, 223 (1989).

2. A.Y. Wong, R.F. Wuerker, J. Sabutis, and R. Suchannek, C. D. Hendricks, and P. Gottlieb, in Proceedings of the International Workshop on Controlled Active Global Environments (CAGE), E. Sindoni and A.Y. Wong (Societa Italiana, di Fisica. Editrice Compositori: Bologna, Italy, 1991), p. 129-142.

3. T.H. Stix, J. Appl. Phys., 66, 5622 (1989); Proceedings of CAGE, ibid, p. 281.

4. R.J. Cicerone, S. Elliot and R.P. Turco, Science, 254, 1191 (1991).

5. G.M.B. Dobson and D.N. Harrison, Proc. R. Soc. London, Ser A, 110, 660 (1926).

6. S. Chapman, Mem. R. Meteorol. Soc. 3, 103 (1930).

7. R.P. Wayne, Chemistry of Atmospheres, (Oxford Univ. Press, Oxford, 1985).

8. M.J. Molina and F.S. Rowland, Nature London, 249, 810 (1974).

9. J. Zinn, C.D. Sutherland, S.N, L.M. Duncan, Atmos. Terr. Phys., 44, 1143-1171 (1982).

10. D. Smith, M.J. Church, Plane. Space Sci., 25, 433 (1977).

11. G. Brasseur and A. Chatel, Annales Geophysicae, I, 173-185 (1983).

12. K.T. Tsang, D.D-M Ho, A.Y. Wong, and R.J. Siverson, Proceedings of CAGE, ibid. p. 143.

13. D. D-M Ho, K.J. Tsang, A.Y. Wong, and R.J. Siverson, Proceedings of CAGE, ibid. p. 157.

14. A.Y. Wong, R.G. Suchannek, and R. Kanner, Phys. Letter. A, 168, 423-428 (1992).

15. L.T. Molina and M.J. Molina, J. Geophys. Res., 91, 14501 (1986).

16. G. Brasseur and S. Solomon, Aeronomy of the Middle Atmosphere, (Reidel, Dordrecht, 1986).

17. D. Sensharma (private communication).

18. E.R. Williams, Sci. Am., 88, (Nov. 1988).

19. K.N. Leung and X. Yao (private communication).

20. P. Hammil and O.B. Toon, Physics Today 24, 34-42 (Dec. 1991).

21. D. Soderman (private communication).

22. S. Darlington (private communication).

23. Y. Chiu (private communication).

24. M. Onda, Y. Morikawa, N. Nagayama and I. Suzuki, <u>Proceedings of the International Aeronautical Sciences</u>, Beijing, PRC, <u>1</u>, (Sept. 20-24, 1992).

25. G.A. Askar'yan, G.M. Batanov, A.E. Barkhudarov, S.I. Gritsinin, E.G. Korchagina, I.A. Kossyi, V.P. Silakov and N.M. Tarasova, JETP Lett., <u>55</u>, 515 (1992).

26. J.M. Dawson, H.C. Kim, D. Arnush, B.D. Fried, R.W. Gould, L.O. Heflinger, C.F. Kennel, T.E. Romesser, A.Y. Wong, R.F. Wuerker, Phys. Rev. Lett. <u>37</u>, 1547 (1976).

27. A.Y. Wong, "Polar Atmospheric Modification and Environmental Mitigation", Bulletin of the American Physical Society, Plasma Physics. <u>37</u>, No. 6, 1403, (November 1992).

28. T.K. Nakamura and A.Y. Wong, "Cyclotron Heating of the Ionospheric Ions by External Pump Waves", EOS, Transactions, American Geophysical Union, <u>73</u>, No. 43, 417 (October 27, 1992).

29. A.Y. Wong and R. G. Brandt, Radio Science, <u>25</u>, No. 6, 1251-1267 (1990).

30. M.J. McCarrick, D.D. Sentman, A.Y. Wong, R.F. Wuerker, B. Chouinard, Radio Science, <u>25</u>, No. 6, 1291-1298 (Nov/Dec. 1990).

31. M.C. Kelly, <u>The Earth's Ionosphere, Plasma Physics and Electrodynamics</u>, (Academic Press, Inc.: San Diego, CA. 1989).

32. T.H. Stix, <u>Waves in Plasmas</u>, AIP, 1992.

33. Y. Chiu, J.M. Cornwall, J.F. Fennell, C.J. Gorney, P.F. Mizera, Space Science Review, 35 211-257, (1983).

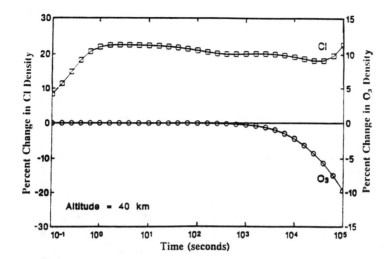

Time evolution of the relative changes of the Cl and O_3 concentrations at an altitude of 40Km after a 23% increase in the ClO concentration.

Figure 1

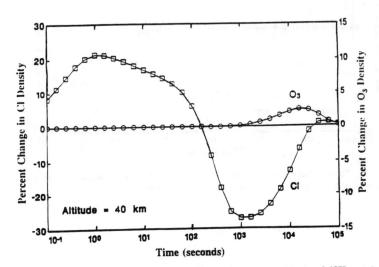

Relative changes of the Cl and O_3 concentration at an altitude of 40Km as function of time after a 23% increase of the ClO concentration and injection of 10^7 electron per cm^3.

Figure 2

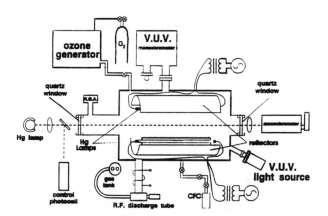

Schematic drawing of the reaction chamber. The chamber has an inner diameter of 1.40 m, and length 4.03 m. The chamber can be evacuated to pressures of the order of 10^{-5} Torr. Two 140 W mercury lamps provide the ultraviolet radiation for studies of photochemical reactions. Absorption of the 253.7 nm line emitted by a low power mercury lamp is used to measure the ozone concentration in the chamber.

Figure 3

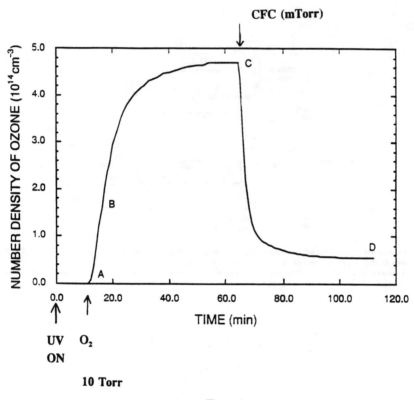

Figure 4

Time evolution of the number density of ozone during the photochemical
generation and subsequent destruction of ozone by trichlorofluoromethane. The
number densities were derived from absorption measurements of the 253.7 nm line.

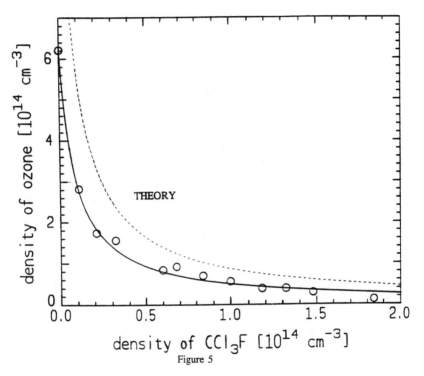

Figure 5

Equilibrium ozone concentration versus density of trichlorofluoromethane. The
ozone concentrations were measured at the equilibria between photochemical ozone
generation and destruction for increasing amounts of Cl_3CF in the reaction
chamber.

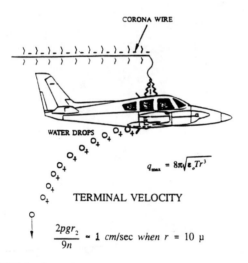

Figure 6

Schematic representation of the injection of electric charges in the atmosphere.

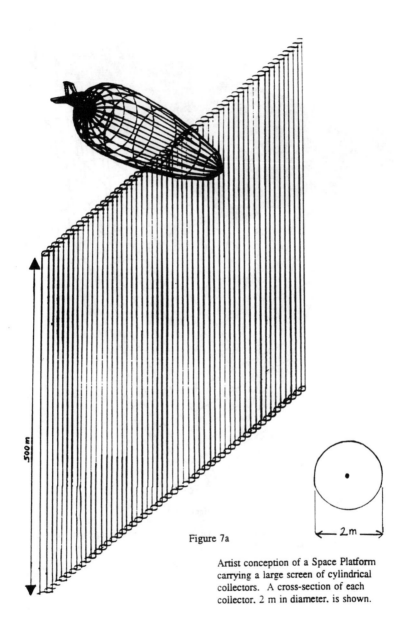

Figure 7a

Artist conception of a Space Platform
carrying a large screen of cylindrical
collectors. A cross-section of each
collector, 2 m in diameter, is shown.

Figure 7b

Artist conception of another version of a Space Platform carrying a cage consisting of screens of opposite bias.

Plasma Heating
by Fast Magnetosonic Waves in Tokamaks

Miklos Porkolab
Department of Physics and
Plasma Fusion Center
Massachusetts Institute of Technology
Cambridge, MA 02139

Invited Lecture presented at the
Stix Symposium
"Advances in Plasma Physics"
May 4, 1992
Princeton, NJ 08540

Work supported by US Department of Energy Contract No. DE-AC02-78ET51013.

Abstract

The fundamental theory of plasma heating by the fast magnetosonic wave in toroidal plasma configurations is reviewed and extended. The particular wave damping processes considered include cyclotron damping at the fundamental ion cyclotron frequency and its harmonics, and electron Landau damping and transit time magnetic pumping (TTMP). The latter processes heat electrons and may also be exploited to drive toroidal plasma currents. The wave absorption and damping decrements are obtained by using Stix's approach, namely by computing the dissipated power, $Re(\bar{J} \cdot \bar{E})$ in terms of the hot plasma dielectric constant (where $\bar{J}$ is the wave induced current). This approach is compared with power absorption calculations from quasi-linear theory, and exact agreement is found for a Maxwellian distribution of particles. Wave absorption in the presence of a small group of energetic particles is also examined for all three types of damping processes. The limitations of theory owing to mode conversion phenomena are indicated. Finally, a brief discussion of recent experimental results is given, verifying the reality of Landau damping of magnetosonic waves by electrons.

I. Introduction

The absorption of the fast magnetosonic wave by electrons in high temperature toka-mak plasmas is of great practical importance because of the feasibility of heating to ignition-like temperatures,[1] as well as driving the toroidal plasma current.[2] An attractive steady state reactor concept could be developed based on fast wave heating and current drive if the bootstrap-current fraction were sufficiently high.[3] From a practical point of view, of considerable importance is the single pass absorption of fast waves by electrons in high temperature plasmas. A single pass absorption of 10% or more is thought to be desirable to ensure unidirectional wave propagation and absorption. The earliest correct calculation of fast wave absorption by electrons was performed by Stix in 1975,[1] who considered only the low frequency limit, $v_{te} \gtrsim \omega/k_\parallel$ where $v_{te} = (2T_e/m_e)^{1/2}$, $\omega/2\pi$ is the wave frequency, and $k_\parallel = \vec{k} \cdot \vec{B}/|\vec{B}|$ is the parallel component of the wave-vector. In such a limit the single pass absorption is rather weak, and the current drive efficiency is also low. As a consequence, one may have to deal with eigen-mode excitation in the toroidal plasma. The low frequency regime ($\omega < \omega_{ci}$) has the advantage of avoiding "parasitic" ion absorption, including that by alpha particles. On the other hand, one may be faced with partial mode conversion into shear Alfven waves[4] where $n_\parallel^2 = S$ ($n_\parallel = ck_\parallel/\omega$ and S is the perpendicular component of the dielectric constant[1]). More recent calculations emphasized the regime $\omega/k_\parallel \gtrsim v_{te}$, $\omega > \omega_{ci}$, so that the theoretically predicted stronger absorption by electrons could be tested in present-day tokamak plasmas.[5-8] Here we shall review the calculations of fast wave absorption on electrons for arbitrary parallel phase velocities.

The dissipation of wave power by electrons is manifested through Landau-type wave-particle resonances, $\omega \simeq k_\parallel v_{e\parallel}$, and the strongest interaction results when the wave resonates with the bulk-electrons, namely $v_{e\parallel} \simeq v_{te}$. It will be shown below that Stix's approach can be followed through even for the case of resonance with bulk electrons if one considers the use of the plasma dispersion function (or Z-function). Similarly to Stix we find[8] that three terms contribute to the damping of this wave: the electron Landau damping being proportional to ImK_{zz} (where ImK is the imaginary part of the hot plasma dielectric tensor), the transit time magnetic pumping (or "TTMP") being proportional to ImK_{yy}, and the cross term, being proportional to ImK_{yz}. Quasi-linear estimates have shown that for $\omega \lesssim 2k_\parallel v_{te}$, in the presence of a fast magnetosonic wave the electron distribution remains nearly a Maxwellian.[9] We shall find that in this case for low frequencies the Landau damping term dominates while the other two terms cancel. However, at finite frequencies ($\omega > \omega_{ci}$) an additional term survives from the cross-term and increases wave damping. We shall also find that significant absorption will result only at finite values of the electron beta, typically $\beta_e \gtrsim 0.01$.

The present calculations were motivated by testing experimentally this kind of heating and current drive concept in the General Atomics DIII-D tokamak.[10] The initial calculations by the author using the Stix-formalism[5,6] was followed by S.C. Chiu et al.[7] who calculated the damping decrement from the determinant of the hot plasma dielectric tensor. These calculations are more direct, however, they do not display the physical importance of the various absorption and wave polarization terms. Therefore, here we shall follow the physically more transparent derivation of taking the ratio of the absorbed power to the power flux which yields the spatial damping rate. For electromagnetic waves, to a good approximation, the power flux is simply given by the Poynting flux. We note that for $\omega > 2k_\parallel v_{te}$, the absorption is generally too weak to be of practical interest.

We shall also generalize these results to the case of two-component plasma, for example when a small fraction of energetic electron component is present. Such a situation may be produced in "synergistic" experiments where in addition to the magnetosonic wave, a lower-hybrid wave may also be present in the plasma. The question of direct fast wave absorption by the energetic electron tail has been raised as an important issue in connection with recent experiments on JET where an enhanced current drive effect was noted.[11]

We shall also consider the case of competing wave absorption process by ions. This is detrimental to current drive or direct heating by electrons, and must be carefully considered. We shall make simple estimates of ion absorption by harmonic ion cyclotron damping, minority absorption and absorption by energetic ions (due to neutral beam injection or alpha particles). Finally, a brief summary of recent experimental results will be presented, supporting evidence of direct absorption of fast waves by electrons in tokamak plasmas. It should be noted that the absorption of fast waves by ions has been tested experimentally over the past decade and a half, and these results are summarized in a companion lecture by J. Hosea.

II. Absorption of Magnetosonic Waves by Electrons

The dispersion relationship of fast waves in the ion cyclotron frequency range may be written in the following form:[1,12]

$$n_\perp^2 = \frac{(n_\parallel^2 - R)(n_\parallel^2 - L)}{S - n_\parallel^2} \tag{1}$$

where in the cold plasma limit the following approximate expressions hold:

$$R = 1 + \sum_i \frac{\omega_{pi}^2}{\omega_{ci}(\omega + \omega_{ci})} \ ,$$

$$L = 1 - \sum_i \frac{\omega_{pi}^2}{\omega_{ci}(\omega - \omega_{ci})} \ ,$$

$$S = 1 - \sum_i \frac{\omega_{pi}^2}{\omega^2 - \omega_{ci}^2} \ ,$$

where $S = (R + L)/2$. Here $\omega_{pi} = (4\pi n_i e^2 Z_i^2/m_i)^{1/2}$ is the angular ion plasma frequency and $\omega_{ci} = eZ_i B/m_i c$ is the angular ion cyclotron frequency. Equation (1) predicts a rather complex behavior for fast wave propagation, especially in the case of multi-ion species plasmas. In addition to Refs.(1,12) an excellent summary of such phenomena can be found in Ref.(13). Here we simply want to point out a few of the salient features of Eq.(1).

The region of $n_\parallel^2 = R$ corresponds to the right hand cut-off layer ($n_\perp^2 = 0$), and the fast wave is evanescent at densities lower than this critical density, n_R. The right hand cut-off layer always exists in the plasma, regardless of the relative value of ω/ω_{ci}. Consequently, the wave has to "tunnel through" an evanscent layer in the plasma periphery and the reflected rf power must be prevented from getting back into the rf source (usually a high power tetrode) by an external tuning (matching) network. The $n_\parallel^2 = L$ layer is also a cut-off layer. In a single ion species plasma, this layer occurs only for $\omega < \omega_{ci}$, and typically it occurs at densities n_L such that $n_L < n_S < n_R$, where n_S is the critical density at the resonance layer where $n_\parallel^2 = S$ and $n_\perp^2 \to \infty$. At the resonance layer finite temperature effects must be included and mode conversion into the kinetic shear Alfven wave will take place. The inhomogeneous magnetic field of a tokamak will only quantitatively change this picture. The presence of the cut-off-resonance-cut-off "triplet" complicates the prediction of rf power flow into the fast magnetosonic wave. If $\omega > \omega_{ci}$ everywhere in the plasma, such a complication does not arise and one only need to be concerned with the right hand cut-off layer. If a minority ion species (or a second majority ion species) is present in the plasma, the mode conversion layer ($n_\parallel^2 = S$) will be affected by the second ion species, and the cyclotron frequency of the lighter ion species (ω_{cm}) will dominate. For example, if $\omega \simeq \omega_{cm}$ near the center of the plasma column, the $n_\parallel^2 = L$ and $n_\parallel^2 = S$ layers will be located on the high field side of the minority species cyclotron resonance layer. The $n_\parallel^2 = R$ layer will remain near the plasma periphery, maintaining the presence of the evanescent layer.

If $\omega > \omega_{cm}$ everywhere in the plasma column, and if we consider regions of fast wave propagation well away from any cut-off or resonance layer, the fast wave dispersion can be approximated fairly well by the following simple relationship:

$$\omega \simeq k_\perp v_A (1 + c^2 k_\parallel^2 / \omega_{pi}^2)^{1/2} , \tag{2}$$

where $v_A \simeq c\omega_{ci}/\omega_{pi}$ is the Alfven speed, and where usually $n_\parallel^2 \ll n_\perp^2$ so that $\omega \simeq k_\perp v_A$. In the discussions below, we shall use only Eq.(2) for the real part of the dispersion relationship when we calculate power absorption.

As shown by Stix, the absorbed power in the plasma can be determined by calculating the dissipated wave power, $Re(\vec{J} \cdot \vec{E})$ which can be written in the form

$$P_{abs} = \frac{-i\omega}{16\pi} \sum_{modes} \vec{E}^* \cdot (\overleftrightarrow{K} - \overleftrightarrow{I}) \cdot \vec{E} + c.c. \tag{3}$$

where the summation is over different modes, $\overleftrightarrow{K}$ is the hot plasma dielectric tensor, $\overleftrightarrow{I}$ is the unit diadic, and c.c. represents the complex conjugate. Thus, the contribution will come from the anti-Hermitian part of $\overleftrightarrow{K}$, and the Hermitian part will cancel. Now we will be interested in Landau-type of resonance of electrons, namely $\omega \simeq k_\parallel v_{e\parallel}$. Examining the hot plasma dielectric tensor,[12] we find that the following terms may contribute:

Landau Damping:

$$K_{zz} = 1 + \frac{1}{k_\parallel^2 \lambda_{De}^2} \left[1 + \zeta_e Z(\zeta_e) \right]; \tag{4a}$$

Transit Time Magnetic Pumping:

$$K_{yy} = 1 + 2k_\perp^2 r_{ce}^2 \frac{\omega_{pe}^2}{\omega^2} \zeta_e Z(\zeta_e); \tag{4b}$$

Cross-Terms:

$$K_{yz} = -K_{zy} = -i \frac{\omega_{pe}^2}{\omega\omega_{ce}} \frac{k_\perp}{k_\parallel} [1 + \zeta_e Z(\zeta_e)]; \tag{4c}$$

Here the cyclotron (harmonic) resonance terms have been neglected. In Eq.(3) we defined $\lambda_{De} = v_{te}/\sqrt{2}\omega_{pe}$, $r_{ce} = v_{te}/\sqrt{2}\omega_{ce}$, $v_{te}^2 = 2T_e/m_e$, $\omega_{pe}^2 = 4\pi n_e e^2/m_e$, $\omega_{ce} = eB/m_e c$, $\zeta_e = \omega/k_\parallel v_{te}$ and $Z(\zeta_e)$ is the Fried-Conte plasma dispersion function. We note that while Landau damping is the result of the force on a charge due to the parallel wave electric field, $(eE_\parallel)$, transit time magnetic pumping results from the force associated with the magnetic moment and the wave magnetic field,[12] $(-\nabla_\parallel(\mu B))$.

The spatial damping decrement is given by the ratio of the absorbed power, P_{abs} and the Poynting flux, $S_\perp$

$$2k_{\perp Im} = P_{abs}/S_\perp \tag{5a}$$

where we take $S \simeq S_\perp \sim cn_x|E_y|^2/8\pi$, we assumed $k_\parallel^2 \ll k_\perp^2$, and $k_\perp = k_x$. Evaluating Eq.(3), the contributions from 4(a-c) are given by

$$P_{abs} = \frac{\omega}{4\pi^{1/2}} \frac{\omega_{pe}^2}{\omega^2} \left[k_\perp^2 r_{ce}^2 |E_y|^2 - \frac{\omega}{\omega_{ce}} \frac{k_\perp}{k_\parallel} |E_z||E_y| + \frac{\omega^2}{k_\parallel^2 v_{te}^2} |E_z|^2 \right] \zeta_e e^{-\zeta_e^2} . \tag{5b}$$

III. Wave Polarization

To proceed, we need to evaluate E_z in terms of E_y, and then substitute for E_z in Eq.(5). This can be carried out with the help of the dielectric matrix equation,[12]

$$\begin{pmatrix} K_{xx} - n_z^2 & K_{xy} & K_{xz} + n_x n_z \\ K_{yx} & K_{yy} - n^2 & K_{yz} \\ K_{zx} + n_x n_z & K_{zy} & K_{zz} - n_x^2 \end{pmatrix} \begin{pmatrix} E_x \\ E_y \\ E_z \end{pmatrix} = 0 \tag{6}$$

which results in three equations relating E_x, E_y, E_z. Here $n^2 = n_z^2 + n_x^2$, and $n_z = n_\parallel = ck_\parallel/\omega$, $n_x = n_\perp = ck_\perp/\omega$ are the parallel and perpendicular components of the index of refraction. Eliminating E_x in favor of E_y and E_z, the following result can be deduced:

$$\frac{E_y}{E_z} = \frac{n_x^2 n_z^2 - (K_{xx} - n_z^2)(K_{zz} - n_x^2)}{-n_x n_z K_{xy} + (K_{xx} - n_z^2)K_{zy}} . \tag{7}$$

We now consider the relative magnitudes of various terms in Eq.(7) for $\omega_{ci} \sim \omega < \omega_{pi}$, i.e. the ion-cyclotron frequency range. The other important approximation is that $n_x^2 < |K_{zz}|$. This usually implies that $\omega^2 < \omega_{LH}^2 \sim \omega_{pi}^2$, the lower-hybrid (ion-plasma) frequency. This follows from the scaling $n_x^2 \sim \omega_{pi}^2/\omega^2$, $K_{zz} \sim (\omega_{pe}^2/\omega^2)$ or $\sim 1/(k_z^2 \lambda_{De}^2)$. Thus in the numerator, $n_x^2 n_z^2$ is negligible for $n_x \sim c/v_A \sim \omega_{pi}/\omega_{ci}$, $n_z = ck_z/\omega \sim c/v_{te}$, and $|K_{xx}| \sim \omega_{pi}^2/\omega_{ci}^2$. The next simplification occurs if we neglect the first term in the denominator, namely $n_x n_z K_{xy}$. As will be shown later, this corresponds to the low frequency, hot plasma limit, namely

$$\frac{\omega^2}{\omega_{pi}^2} \ll \frac{T_e}{m_e c^2} \tag{8}$$

in which case $E_y/E_z \simeq -K_{zz}/K_{zy}$, or as shown by Stix,[1]

$$\frac{E_z}{E_y} = -\frac{ik_\perp k_\parallel v_{te}^2}{2\omega\omega_{ce}} \; .$$

(9)

This result is valid for arbitrary values of $\omega/k_\parallel v_{te}$, (as long as the unity term in K_{zz} can be ignored). This results from the fact that $[1 + \zeta_e Z(\zeta_e)]$ cancels in the ratio of K_{zz}/K_{zy}. Note that for $\omega \sim k_\parallel v_{te}$, $|E_z/E_y| \sim k_\perp r_{ce}/2 \sim 10^{-3}$ and therefore the electron absorption will be relatively weak. Using Eq.(9) in (5b) results in a cancellation of the first and second terms (i.e., TTMP and the cross term cancel) and only the third term, namely Landau damping survives. As noted by Stix,[1] its magnitude is 1/2 that of TTMP. Thus, the damping of the fast magnetosonic wave in a Maxwellian plasma at low frequencies and high temperatures is entirely due to Landau damping for arbitrary values of the wave phase velocity.

IV. Damping Rate

Dividing the last term of Eq.(5) with S and using Eq.(9), we obtain for the spatial damping rate

$$2k_{\perp Im} = k_{\perp Re}\left(\frac{\pi^{1/2}}{2}\right)\beta_e \zeta_e \exp(-\zeta_e^2)$$

(10)

where $\beta_e = 8\pi n_e T_e/B^2$ is the electron beta. Replacing $k_{\perp Re} \sim \omega/v_A$, we get

$$2k_{\perp Im} = \frac{\pi^{1/2}}{2}\frac{\omega}{\omega_{ci}}\frac{\omega_{pi}}{c}\beta_e \zeta_e \exp(-\zeta_e^2)$$

(11)

so that for $\zeta_e \sim 1$, single pass absorption is proportional to ω, $n^{3/2}$, T_e and B^{-3}. We also note that the maximum absorption occurs for $\zeta_e \sim 0.7$. For example, for present day machines, $T_{eo} \simeq 6.0$ keV, $B_o \simeq 2.0$ T, $n_o \sim 5 \times 10^{19}\text{m}^{-3}$, $\beta_e \sim 0.03$, $f = 76$ MHz in D plasma, $\zeta_e \sim 0.7$, $\Delta x \sim a/2 \sim 0.50$ m, $2k_\perp \Delta x \simeq 0.62$, and the single pass absorption is $[1 - \exp(-2k_{\perp Im}\Delta x)] \simeq 0.47$. This may be a typical achievable value in the DIII-D tokamak. The required parallel wavelength at the antenna would be $\lambda_\parallel \simeq 60$ cm which is very reasonable (toroidal wave propagation effects would reduce this to $\lambda_\parallel \simeq 44$ cm near the center of the plasma where $\zeta_e \simeq 0.7$).

Now we consider the more general result in Eq.(7), namely retain the first term in the denominator (i.e. do not assume Eq.(8)). Taking the inverse of Eq.(7), we obtain

$$\frac{E_z}{E_y} = -i\frac{k_\perp k_\| v_{te}^2}{2\omega\omega_{ce}} + \frac{K_{xy}n_x n_z}{(K_{xx} - n_z^2)K_{zz}} \cdot \tag{12}$$

Now we find that as before, in Eq.(5) the TTMP term cancels with the cross term for the first term of Eq.(12). However, the second term of Eq.(12) survives with the cross-term and adds to the Landau damping term, increasing its effectiveness at higher frequencies. The net damping decrement is obtained by combining Eqs.(5) and (12), and upon dividing by the Poynting flux we obtain

$$2k_{\perp Im} = k_\perp Re(\frac{\pi^{1/2}}{2})\beta_e\zeta_e e^{-\zeta_e^2}[1 + \frac{1}{\alpha^2}] \tag{13}$$

where the surviving cross-term gives

$$\alpha = \frac{T_e}{m_i c^2}(\frac{\omega^2 - \omega_{ci}^2}{\omega_{pi}^2})(S - n_\|^2)|K_{zz}| \tag{14}$$

where

$$S = 1 - \sum_j \omega_{pj}^2/(\omega^2 - \omega_{cj}^2)$$

is the cold plasma limit of K_{xx}. Note that in Eq.(14) the absolute value of K_{zz} is to be taken which requires special attention if $\zeta_e \simeq 0(1)$. For $|S| \gg n_\|^2$, and $|S| \gg 1$, Eq.(14) can be written in the following form:

$$\frac{1}{\alpha^2} = \left(\frac{m_e c^2}{T_e}\right)^2 \frac{\omega^4}{4\omega_{pi}^4} \frac{1}{\zeta_e^4|1 + \zeta_e Z(\zeta)|^2} \tag{15}$$

In the cold plasma limit ($\zeta_e^2 \gg 1$), Eq.(15) reduces to

$$\frac{1}{\alpha^2} \simeq \left(\frac{m_e c^2}{T_e}\right)^2 \frac{\omega^4}{\omega_{pi}^4}, \tag{16a}$$

whereas in the hot plasma limit ($\zeta_e^2 \ll 1$), Eq.(15) becomes

$$\frac{1}{\alpha^2} \simeq \left(\frac{m_e c^2}{T_e}\right)^2 \frac{\omega^4}{4\omega_{pi}^4} \frac{1}{\zeta_e^4(1 - 0.86\zeta_e^2)} \simeq \left(\frac{\omega^4}{\omega_{pi}^4}\right)n_\|^4. \tag{16b}$$

The result Eq.(16a) has been noted previously by Moreau et al.[9] For example, for $\zeta_e \simeq 0.7$ and the previously listed plasma parameters $1/\alpha^2 \simeq 0.31$. However, lowering the electron temperature from $T_e \simeq 6.0$ keV to $T_e \simeq 3.0$ keV increases $1/\alpha^2$ to unity. At higher phase velocities ($\zeta_e \gtrsim 1$), $1/\alpha^2$ is less significant for $T_e \gtrsim 3.0$ keV. In Fig. 1 we give a numerically evaluated plot of Eq.(15). Note the dramatic increase of $1/\alpha^2$ for $\zeta_e \lesssim 1$, reflecting the ζ_e^{-4} dependence of $1/\alpha^2$ in this limit.

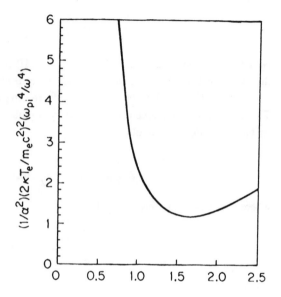

Fig. 1 The normalized value of $1/\alpha^2$ as a function of ζ_e.

V. Two Component Plasma with a High Energy Electron Tail

In recent experiments in JET a synergism between high phase velocity fast waves and a pre-formed electron tail, driven by lower-hybrid waves, has been discovered.[11] Similar phenomena might be expected to occur if fast waves were launched into a runaway dominated discharge. Here of interest is a range of phase velocities such that

$$v_t \ll \omega/k_\| \simeq v_{h\|} \tag{17}$$

where $v_{h\|} = (2T_{h\|}/m_e)^{1/2}$ is the effective mean velocity of the hot electrons, while v_t is that of the cold electrons. Such high phase velocity waves can be launched by a "monopole" type phasing of the fast wave antenna current straps (i.e., 2 or more current straps fed in phase from the transmitters). For example, in a typical lower-hybrid current driven discharge, $T_{h\|} \simeq 100 - 500$ keV, $T_{h\perp} \sim 70$ keV, $T_e \sim 1 - 5$ keV, and $0.005 \lesssim n_h/n_e \lesssim 0.01$, where n_h is the density of hot electrons.[14] Here we shall assume that the hot component is also characterized by a Maxwellian:

$$f_e(\vec{v}) = \frac{n_h}{\pi^{3/2}v_{h\perp}^2 v_{h\parallel}} e^{-\left(\frac{v_{\parallel}^2}{v_{h\parallel}^2} + \frac{v_{\perp}^2}{v_{h\perp}^2}\right)} \tag{18}$$

We can repeat the previous calculations for an anisotropic plasma. The polarization will be mainly determined by the cold plasma, whereas the absorption will be on the hot plasma component. In the limit where the effective $1/\alpha^2$ is negligible, the result of the calculations is as follows:[8]

$$2k_{\perp Im} = k_{\perp Re}\left(\frac{\pi^{1/2}}{2}\right)\beta_{h\perp}\frac{T_{h\perp}}{T_{h\parallel}}\zeta_h e^{-\zeta_h^2}\left[2\left(1 - \frac{T_{\perp}}{T_{h\perp}}\right) + \frac{T_{\perp}^2}{T_{h\perp}^2}\right] \tag{19}$$

where the first term in the bracket corresponds to TTMP, the second term is the cross-term and the third term is the Landau damping term. Here $T_{\perp}$ designates the temperature of the cold plasma component. We have designated the beta of the hot perpendicular component by

$$\beta_{h\perp} = \frac{8\pi n_{eh}T_{h\perp}}{B^2} \tag{20}$$

and $\zeta_h = \omega/k_{\parallel}v_{h\parallel}$. Since in the present case $T_{\perp} \ll T_{h\perp}$, we see that the cross term and the Landau damping terms are negligible as compared to TTMP (the first term in the bracket). Furthermore, if $T_{h\perp} \ll T_{h\parallel}$, as is often the case, for $\zeta_{h\parallel} \simeq 1$, $\beta_{h\perp} = \beta_{bulk}$, the absorption is reduced as compared to the bulk absorption case when $\zeta_{bulk} \simeq 1$. Substituting in typical numbers from the JET experiments we find that Eq.(19) predicts very weak single pass absorption, of the order of one percent. Therefore, we expect a rather weak effect on the overall current drive efficiency, in disagreement with experimental results. These experiments were complicated by the fact that a minority component and ion cyclotron resonance layer were also present. This would introduce an ion Bernstein wave (IBW) mode-conversion layer and diffusion of fast electrons by these waves would have to be invoked to explain the results.[15]

Let us now introduce the finite frequency correction ($1/\alpha^2$ term) into the wave polarization while maintaining absorption on the hot tail (i.e., $\zeta_{h\parallel} \simeq 1$, $\zeta_e \gg 1$, where $\zeta_e = \omega/k_{\parallel}v_{t\parallel}$ designates the bulk plasma). Then it is straightforward to show that the polarization will be determined by the bulk plasma. Carrying out the algebra, Eq.(19) will be modified by inclusion of two additional terms:

$$2k_{\perp Im} = k_{\perp Re}\left(\frac{\pi^{1/2}}{2}\right)\beta_{h\perp}\frac{T_{h\perp}}{T_{h\parallel}}\zeta_h e^{-\zeta_h^2}\left[2(1 - \frac{T_{\perp}}{T_{h\perp}}) + \frac{T_{\perp}^2}{T_{h\perp}^2}\right.$$

$$+ \frac{2}{\alpha} \frac{T_\perp}{T_{\perp h}} \left(1 - \frac{T_\perp}{T_{\parallel h}} \right) + \frac{1}{\alpha^2} \frac{T_\perp^2}{T_{\perp h}^2} \Bigg] \tag{21}$$

where we assumed that $K_{zzRe}/|K_{zz}| \simeq 1$, and in particular, $|K_{zzRe}| \simeq \omega_{pe}^2/\omega^2$ due to the "fluid" approximation on the bulk plasma. Note that for $T = T_h$, Eq.(21) will reduce to Eq.(13) as expected. Furthermore, for $\alpha \to \infty$, Eq.(21) reduces to Eq.(19). In the ICRF regime, for $T_{\perp h} \gg T_\perp$, the first term (TTMP) will remain dominant. However at higher frequencies ($\omega_{ci} \ll \omega < \omega_{LH}$) the last term could dominate. We note that at $\omega \sim \omega_{LH}$, Eq.(21) is not strictly valid since some of the approximations may not hold (in particular, as discussed earlier the approximation $n_x^2 \ll K_{zz}$ may not hold when $\omega \sim \omega_{LH}$). We note that a result similar to Eq.(21) has been derived recently by Moroz et al.[16]

VI. Absorption of Magnetosonic Waves on Ion Cyclotron Harmonics

One of the competing mechanisms for the absorption of fast magnetosonic waves is absorption on ions near the fundamental, or the harmonics of the ion cyclotron frequency.[7] This may occur on bulk ions, or fast ions due to simultaneous neutral beam injection, or even on alpha particles in a reactor grade plasma. From the current drive point of view this must be regarded as "parasitic" absorption since it removes effective power from electrons which drive the current. The ion absorption can occur by means of direct ion absorption, or by means of mode conversion into an ion Bernstein wave (IBW). The latter process may dominate if $n_\parallel \simeq 0$. Here we shall assume that $n_\parallel$ is finite (and in particular, $\omega/k_\parallel v_{te} \sim 1$ or $n_\parallel \simeq v_{te}/c$), and that the fast wave power density is not high enough to distort the initial Maxwellian distribution of ions. When this condition is violated the situation becomes considerably more complex.[1]

Ion cyclotron harmonic absorption in the limit of near-perpendicular propagation may be obtained from the general result Eq.(3). After a considerable amount of algebra, for $b_i \ll 1$ and $\omega \simeq \ell \omega_{ci}$, Eq.(3) reduces to

$$P = \frac{\omega_{pi}^2}{16\pi^{1/2}} \frac{b_i^{(\ell-1)}}{k_\parallel v_{ti}} \frac{\ell}{(\ell-1)! 2^{(\ell-1)}} |E_+|^2 \exp\left[-\frac{(\omega - \ell\omega_{ci})^2}{k_\parallel^2 v_{ti}^2} \right] \tag{22}$$

where we retained hot plasma terms with E_x, E_y, K_{xx}, K_{yy}, K_{xy} and K_{yx}. Here $E_+ = E_x + iE_y$ is the left hand polarized component of the wave electric field, and $b_i = k_\perp^2 r_{ci}^2$ is the finite ion Larmor radius factor ($r_{ci} = v_{ti}/2^{1/2}\omega_{ci}$). This formula is valid for $b_i \ll 1$, which is usually satisfied for not too high harmonics since $k_\perp \sim \omega/v_A$ so that $k_\perp^2 r_{ci}^2 \sim \ell^2 \beta_i/2$.

In the simplest case, wave polarization is obtained from the cold plasma dielectric tensor Eq.(6) where the following dielectric constant terms are used:[12]

$$K_{xx} = K_{yy} = S = 1 - \sum_j \frac{\omega_{pj}^2}{\omega^2 - \Omega_j^2} \simeq -\frac{\omega_{pi}^2}{\omega^2 - \omega_{ci}^2} \tag{23}$$

$$K_{xy} = -iD = -i \sum_j \frac{\omega_{pj}^2 \Omega_j}{(\omega^2 - \Omega_j^2)\omega} \simeq -\frac{i\omega_{pi}^2}{\omega^2 - \omega_{ci}^2}\left(\frac{\omega}{\omega_{ci}}\right) \tag{24}$$

where the last equality in each line is valid in the limit $\omega \sim \ell\omega_{ci} \ll \omega_{pi}$, and $1 \ll \omega_{pi}^2/\omega_{ci}^2$. We also ignore K_{zz} and E_z. Then the polarization is obtained by combining the first two rows of Eq.(6):

$$\left|\frac{E_+}{E_y}\right|^2 \simeq \frac{(\frac{\omega}{\omega_{ci}} - 1)^2(\frac{\omega}{\omega_{ci}} - \cos^2\theta)^2}{|\frac{\omega}{\omega_{ci}}(1 + \cos^2\theta)|^2} \Rightarrow (\ell - 1)^2 \tag{25}$$

where $\ell = \omega/\omega_{ci}$, $\cos\theta \simeq k_\parallel/k$, and the last limit is valid for $\cos\theta \to 0$. Equation (25) predicts the well known result that at $\omega = \omega_{ci}$, $E_+ \to 0$ since the ions shield out the left-hand polarized component of the electric field (i.e., the magnetosonic wave becomes purely right-hand polarized for perpendicular propagation). As is well known,[1] if strong ion absorption is desired, this problem may be remedied by injecting a minority ion species into the plasma, typically a few percent of hydrogen or helium-3 ions into a deuterium majority plasma. Thus ion cyclotron absorption becomes effective again at $\omega = \omega_{cm}$, where the subscript m designates the minority species, since $E_+(\omega \neq \omega_{cM}) \neq 0$ (where M designates the usually heavier ion species). For $\omega = 2\omega_{CD} = \omega_{CH}$, $|E_+/E_y|^2 = 1$, while for a He-3 minority resonance in a deuterium plasma $\omega = (4/3)\omega_{CD}$, and $|E_+/E_y|^2 = 1/9$. As a consequence, in a deuterium plasma He-3 minority absorption is weaker than that due to H minority.

We can obtain the damping by integrating the power absorbed across a cyclotron harmonic resonance layer in a radially inhomogeneous magnetic field, and divide the absorbed power by the Poynting flux, $S_\perp = (c/8\pi)n_\perp|E_y|^2$. The dominant factor in the integral comes from the exponential factor,

$$\frac{\omega - \ell\omega_{ci}(x)}{k_\parallel v_{ti}} \simeq \frac{x}{2^{\frac{1}{2}}\Delta} \tag{26}$$

where we wrote $\omega_{ci}(x) \simeq \omega_{ci}(1 + x/R)$, $\omega \simeq \ell\omega_{ci}$, R is the tokamak major radius and $\Delta \equiv k_\parallel v_{ti} R/2^{1/2}\omega$. Without loss of generality, the limits of integration may be extended to infinity and we have

$$2 \int_{-\infty}^{\infty} k_{Im}(x)dx \propto \frac{1}{2^{\frac{1}{2}}\Delta} \int_{-\infty}^{\infty} \exp(-x^2/2\Delta^2) = \pi^{1/2} \ ,$$

so that the spatial damping decrement $2 \int k_{Im}(x)dx \equiv 2\eta$ becomes

$$2\eta = \frac{\pi}{2} \frac{\omega_{pi}}{c} \frac{R\beta_i^{(\ell-1)} \ell^{2(\ell-1)}}{(\ell-1)! 2^{2(\ell-1)}} \frac{|E_+|^2}{|E_y|^2} \ . \tag{27a}$$

Here we assumed $\omega \simeq k_\perp v_A$ (neglecting $k_\parallel$) so that $k_\perp \simeq \ell \omega_{pi}/c$, and $\beta_i = 8\pi n_i T_i/B^2$ is the ion beta. This result can be evaluated easily for polarization in the cold plasma limit, in particular, combining Eqs.(25) and (27) gives

$$2\eta = \frac{\pi}{2} \frac{\omega_{pi}}{c} R(\frac{\ell^2 \beta_i}{4})^{(\ell-1)} \frac{(\ell-1)}{(\ell-2)!} \tag{27b}$$

where we rearranged some of the numerical factors. The transmission coefficient is given by

$$T = e^{-2\eta} \ , \tag{28a}$$

and absorption is given by

$$A = 1 - T = 1 - e^{-2\eta} \ . \tag{28b}$$

In particular, we have the following results for $\ell = 1$ to 4:

Table I

ℓ	2η
1	0
2	$R\pi\omega_{pi}\beta_i/2c$
3	$R\pi\omega_{pi}\beta^2(81/16c)$
4	$R\pi\omega_{pi}\beta^3(48/c)$

We recall once more that for the low harmonic numbers which are of practical interest we assumed $\ell^2\beta_i/2 \ll 1$.

A comparison of Eqs.(13) and (27) give the single-pass electron to ion cyclotron harmonic absorption ratio, $r = 2ak_{\perp Im}^{avg}/2\eta_{ion}$, or

$$r = (\frac{2}{\pi})^{1/2}(\frac{a}{R})\frac{\ell(\ell-2)!}{(\ell-1)}\frac{\beta_e <\theta>}{(\ell^2\beta_i/4)^{(\ell-1)}} , \qquad (29)$$

where a is the minor radius, and we assumed that the effective "average" absorption distance is the hot core region of the plasma column, or $<\Delta r>_{eff}\simeq d/2 = a$, with d being the plasma diameter. In this region we take

$$<\theta> =< \zeta_e \exp(-\zeta_e^2)(1+\alpha^{-2}) >_{avg} ,$$

where an optimized value of $<\theta> \simeq 0.5$ is assumed (i.e., $\zeta_e \lesssim 1.0$). Thus, for $a/R \simeq 1/3$, $\ell = 2$ we get $r \simeq (\beta_e/4\beta_i)$ so that absorption on ions dominates for $\beta_i \simeq \beta_e$. However, for $\ell \geq 3$, $\beta_e \simeq \beta_i \lesssim 0.10$, electron absorption dominates over ion cyclotron harmonic absorption. Thus, for effective fast wave current drive we should use $\omega \gtrsim 4\omega_{ci}$ or $\omega \lesssim \omega_{ci}$. Note that additional ion absorption mechanisms due to minority species (H or He-3), neutral beam particles or alpha particles need to be considered.

VII. Mode Conversion Regime

We can further improve on this theory by including the hot plasma contribution to the polarization calculation. In particular, by retaining the hot plasma contribution to K_{xx} and K_{xy}, we have

$$\frac{E_x}{E_y} = -\frac{K_{xy}}{K_{xx}} = \frac{-i\ell A - iBZ_{Im}}{A + BZ_{Im}} \qquad (30)$$

where

$$A = -\omega_{pi}^2/(\omega^2 - \omega_{ci}^2)$$

$$B = \frac{b_i^{\ell-1}\omega_{pi}^2\ell^2}{2k_{\parallel}v_{ti}\omega 2^{(\ell-1)}\ell!}$$

and

$$Z_{Im} = \sqrt{\pi}\exp(-(\omega - \ell\omega_{ci})^2/k_{\parallel}^2v_{ti}^2).$$

Thus, we can evaluate $|E_+/E_y|^2$ and obtain the following result for the polarization:

$$\frac{|E_+|^2}{|E_y|^2} = \frac{(\ell-1)^2}{1+\sigma_\ell^2} \qquad (31a)$$

where we assumed $\ell \gtrsim 2$, and

$$\sigma_\ell^2 \simeq \pi^{1/2}\left(\frac{\ell^2\beta_i}{4}\right)^{\ell-1}\frac{(\ell^2-1)}{2\ell!}\zeta_{oi}F\ , \qquad (31b)$$

where $\zeta_{oi} = \omega/k_{\parallel}v_{ti}$ and $F \simeq \exp(-2\zeta_{-\ell}^2)$. Thus, to correct for polarization effects, including hot plasma corrections, we simply divide Eq. 27(b) by $(1 + \sigma_\ell^2)$, where for simplicity we take $F \Rightarrow 1$ (more accurately, we should repeat the intergration over x but the final result remains the same as taking $F \Rightarrow 1$).

Note that σ_ℓ corresponds to the ratio of δ, the separation between the cyclotron resonance layer and the mode conversion layer, to $\Delta \propto k_{\parallel}v_{ti}R/\omega$, the width of the cyclotron resonance layer. The "cold plasma" result is obtained in the limit $\sigma_\ell^2 \ll 1$. For example, for $\ell = 2, \delta \simeq \beta_i R/2$ and $\delta/\Delta \simeq \beta_i\zeta_{oi} \sim \sigma_2$. For $\sigma^2 \ll 1$ cyclotron harmonic absorption dominates, and if $2\eta > 1$, we have effective absorption on ions. In the opposite limit, $\sigma_\ell > 1$, we have effective "depolarization," and we end up in the mode-conversion regime (from the fast wave to ion Bernstein wave or IBW). It has been shown[18] that in the mode conversion regime for low field side launch the transmission (T), reflection (R) and mode conversion (M) coefficients are given by the Budden factors

$$T = \exp(-2\eta) \tag{32a}$$

$$R = (1 - T)^2 \tag{32b}$$

$$M = T(1 - T) \tag{32c}$$

so that $R + T + M = 1$.

The results for high-field side launch are

$$T = e^{-2\eta} \tag{33a}$$

$$R = 0 \tag{33b}$$

$$M = 1 - T \tag{33c}$$

so that effective mode conversion takes place since waves arrive at this layer first. The fate of the mode converted IBW is somewhat complicated. On the "midplane" it may simply convect out of the plasma, while off the midplane it may be absorbed by electrons.

VIII. Minority Absorption Regime

It is straightforward to include absorption by minority ion species in the previous theory. The absorbed power is

$$P_m = \frac{\omega_{pm}^2}{16\pi^{1/2}} \frac{|E_+|^2}{k_\parallel v_{tm}} \exp\left[-\frac{(\omega - \omega_{cm}^2)}{k_\parallel^2 v_{tm}^2}\right] \tag{34}$$

where m is the minority species. Integrating across the resonance layer as before, the damping decrement is given by

$$2\eta = \frac{\pi}{2}\frac{\omega_{PM}}{c}\frac{n_m}{n_M}\frac{Z_m}{Z_M}R|E_+|^2/|E_y|^2 \tag{35}$$

where R is the major radius, M designates the majority species, ω_{PM} is the majority angular ion plasma frequency, Z is the ion charge, and n_m/n_M is the minority to majority density ratio. We can again carry out the hot plasma polarization calculations and obtain

$$\frac{|E_1^+|^2}{|E_y|^2} = \frac{1}{1+\sigma_1^2} \tag{36}$$

where

$$\sigma_1^2 = \frac{\pi}{4}\left(\frac{n_m}{n_M}\frac{M}{m}\frac{Z_m^2}{Z_M^2}\right)^2\left(1 - \frac{\omega_{cM}^2}{\omega^2}\right)^2\left(\frac{\omega}{k_\parallel v_{ti}}\right)^2 \tag{37}$$

again separates the ion absorption regime ($\sigma_1^2 \ll 1$) to the mode-conversion regime ($\sigma_1^2 > 1$). Here M/m is the majority to minority ion mass ratio, and ω_{cM} is the majority ion cyclotron frequency (here $\delta \sim Rn_m/n_M$ and $\Delta \sim k_\parallel v_{tm}R/\omega$ and the ratio gives σ_1). We note that for H^+ minority, D^+ majority ions $k_\perp \sim 2\omega_{pM}/c$, and

$$2\eta = \frac{\pi}{2}\frac{\omega_{pM}}{c}\frac{n_m}{n_M}R \; . \tag{38}$$

Thus, the role of β_i (second harmonic deuterium absorption) has been replaced by n_m/n_M, the fraction of minority ions. If $\beta_M < n_m/n_M$, minority absorption dominates. Note that minority absorption is very effective even in a relatively cold plasma when β_M may be low.

IX. Absorption of Magnetosonic Waves at the Fundamental Majority Ion Cyclotron Resonance ($\ell = 1$)

Although in the cold plasma limit for perpendicular propagation we found no absorption, by introducing finite values of $k_\parallel$, $E_+ \neq 0$ if the ion temperature is sufficiently high. It is a simple matter to show that by including Doppler-shifts due to finite $k_\parallel v_{ti}$, the left hand polarized component of the fast wave is finite and Eq.(31a) would be replaced with

$$|E_+|^2/|E_y|^2 = \frac{2}{\pi}k_\parallel^2 r_{ci}^2 \; . \tag{39}$$

where r_{ci} is the ion gyro-radius $(b_i^{1/2}/k_\perp)$. Combining this with the power absorption formula at $\omega = \omega_{ci}$ we obtain

$$2\eta = \frac{\omega_{pi}}{c} R k_\parallel^2 r_{CM}^2 \ . \tag{40}$$

Noting that $k_\parallel^2 r_{CM}^2 = k_\parallel^2 v_{ti}^2/2\omega^2 = n_\parallel^2 T_i/m_i c^2$ for $\omega = \omega_{CM}$, the finite absorption depends on Doppler shift. We note that majority ion cyclotron resonance absorption is weaker than harmonic ion cyclotron absorption by the ratio $(2/\pi)k_\parallel^2 r_{CM}^2/\beta_M \simeq (n_\parallel^2 \omega_{ci}^2/\omega_{pi}^2 \pi) \ll 1$. Similar results apply in comparison with minority absorption or with electron absorption. It should be noted, however, that this treatment ignores potential difficulties with mode conversion into shear Alfven waves on the high field side of the resonance.[4]

X. Harmonic Ion Cyclotron Absorption by High Energy Ions

Absorption of magnetosonic waves by hot ions may be of importance when simultaneous neutral beam injection is taking place, or when the rf power is strong enough to form an energetic ion tail by quasi-linear diffusion.[1] The ions have a slowing-down energy distribution which is best modelled by Monte-Carlo techniques in the case of neutral beam injection. As shown by simulations, a typical distribution function may be characterized in an approximate way by a Maxwellian with energies $T_{h\parallel} \sim \varepsilon_{bmax}/3$, $T_{h\perp} \sim \varepsilon_{bmax}/4$, and an ion population of a few percent of the bulk ions. If a harmonic cyclotron resonance layer is present in the plasma, significant wave power loss to the beam ions may result, while the beam ions would be accelerated to higher energies. This may be beneficial to beam penetration near the center of the plasma while detrimental near the edge (depending on the location of the cyclotron harmonic layer). In any case, it would be detrimental to driving plasma currents with the wave. It should be noted that such beam acceleration has been found in recent 3rd harmonic resonance experiments.[19]

We can easily repeat the previous calculations, with special care given to absorption by the hot ions while using the bulk plasma parameters for dispersion and polarization. The result of the calculation is

$$2\eta = \frac{\pi}{2} \frac{\omega_{pb} R}{c} (\frac{\ell^2 \beta_{h\perp}}{4})^{(\ell-1)} (\frac{n_b}{n_h})^{(\ell-2)} \frac{(\ell-1)}{(\ell-2)!} \ , \tag{41}$$

where sub-b designates the bulk (background) ions, sub-h designates the hot ions, $\beta_{h\perp} = 8\pi n_h T_{h\perp}/B^2$ is the hot ion beta perpendicular component, and ℓ is the harmonic number

($\ell = \omega/\omega_{ci}$). Note again, that this result is valid only for $k_\perp^2 r_{ch}^2 = (\ell^2/2)(n_b/n_h)\beta_h \ll 1$ and $\ell \geq 2$. We see that formula (41) has some interesting dependence on harmonic numbers. Comparing Eqs.(27b) and (41), we find for the ratio of cyclotron harmonic absorption on hot (beam) ions versus that on bulk ions is

$$r = \left(\frac{\beta_{h\perp}}{\beta_b}\right)^{\ell-1}\left(\frac{n_b}{n_h}\right)^{\ell-2}. \tag{42}$$

It is interesting to note that for $\ell = 2$, $r = \beta_h/\beta_b$ and absorption on hot ions may be comparable to, or less than that on the bulk ions. However, for $\ell \geq 3$, even if $\beta_h/\beta_b \lesssim 1$, absorption on the hot ions may well dominate if $n_b \gg n_h$.

XI. Quasi-Linear Theory

Another way to consider absorption of waves in the plasma is by means of quasi-linear theory. This formalism has the advantage of easy generalization to non-Maxwellian distributions, including that created by the incident rf waves themselves. Although the waves may be coherent, the particles transiting them lose phase memory as they pass around the torus hundreds of times and experience rare collisions.[1] To obtain the true distribution function of a species of charged particles, one must solve a Fokker-Planck equation, including quasi-linear diffusion and collisional drag and diffusion of the form[1]

$$\frac{\partial f}{\partial t} = \frac{\partial f}{\partial t}|_{QL} - \nabla_{\vec{v}}(< \Delta \vec{v} > f) + \frac{1}{2}\nabla_{\vec{v}} \cdot [\nabla_{\vec{v}} \cdot (< \nabla \vec{v}\nabla \vec{v} > f)] \tag{43a}$$

where the 2nd and 3rd terms on the RHS correspond to collisional tems and

$$\frac{\partial f}{\partial t}|_{QL} = \frac{\partial}{\partial \vec{v}} \cdot \overleftrightarrow{D}_{QL} \cdot \frac{\partial f}{\partial \vec{v}} \tag{43b}$$

is the quasi-linear term. In general this is a difficult problem which has been solved only in a few instances. For example the case of a minority species distribution function in the presence of ICRF heating and collisional drag has been determined by Stix.[1] The experimental verification of this theory has been one of the triumphs of ICRF experiments on tokamaks during the past decade, and it will be discussed by other authors in these Proceedings. In the steady state the result is the characterization of a high energy minority tail by an effective temperature[1]

$$\frac{1}{T_{eff}} = \frac{1}{T_e(1+\zeta)}\left[1 + \frac{R_j(T_e - T_j + \zeta T_e)}{T_j(1 + R_j + \zeta)}\frac{1}{1 + (E/E_j)^{3/2}}\right] \tag{44}$$

where

$$R_j = \frac{n_j Z_j^2}{n_e} \left(\frac{v_{te}}{v_{tj}} \right),$$

$$\zeta = \frac{m \langle P \rangle v_{te}}{8\pi^{1/2} n_e n Z^2 e^4 \ell n \Lambda},$$

and $\langle P \rangle$ is the average power per unit volume deposited. Here the majority ion species is characterized by density n_j, temperature T_j, charge eZ_j, thermal speed $v_{tj} = (2T_j/m_j)^{1/2}$; electrons are characterized by density n_e, and thermal speed v_{te}; the minority species being accelerated by cyclotron resonance are designated by n, m, v, Z and $E = mv^2/2$. We note that for $\zeta \simeq 0$ the minority ion species is characterized by a temperature close to that of the majority ion temperature, whereas for $E \gg E_j(\zeta)$, $T_{eff} \simeq T_e(1 + \zeta)$ and ion acceleration is entirely balanced by electron drag.

Let us now ignore collisions and discuss the quasi-linear diffusion term in the presence of rf waves. In particular, the power absorbed by charged particles can be calculated as follows:

$$P = \int d^3v \frac{mv^2}{2} \left(\frac{\partial f_o}{\partial t} \right)_{QL}. \tag{45}$$

The quasi-linear evolution of the distribution function in a magnetized plasma has been given by Kennel and Engelmann nearly three decades ago[17] and it may be written in the following form:

$$\left(\frac{\partial f}{\partial t} \right)_{QL} = V \stackrel{lim}{\rightarrow} \infty \sum_{\ell} \frac{\pi Z^2 e^2}{m^2} \int \frac{d^3k}{(2\pi)^3 V} \mathcal{L} v_\perp \delta(\omega - \ell\Omega - k_\parallel v_\parallel) |\Theta_{\ell,k}|^2 v_\perp \mathcal{L} f \tag{46}$$

where

$$\mathcal{L} = \left(1 - \frac{k_\parallel v_\parallel}{\omega} \right) \frac{1}{v_\perp} \frac{\partial}{\partial v_\perp} + \frac{k_\parallel}{\omega} \frac{\partial}{\partial v_\parallel}$$

$$\Theta_{\ell,k} = \frac{1}{2} e^{+i\psi} (E_x - iE_y)_k J_{\ell+1} \left(\frac{k_\perp v_\perp}{\Omega} \right)$$

$$+ \frac{1}{2} e^{-i\psi} (E_x + iE_y)_k J_{\ell-1} \left(\frac{k_\perp v_\perp}{\Omega} \right)$$

$$+ \frac{v_\parallel}{v_\perp} E_{zk} J_\ell \left(\frac{k_\perp v_\perp}{\Omega} \right)$$

where $\vec{k} = k_\perp \cos\psi \hat{x} + k_\perp \sin\psi \hat{y} + k_\parallel \hat{z}$. Here V is the plasma volume, $\vec{E}_k$ are the Fourier amplitudes of the electric field and ψ is the angle between the $\hat{x}$ direction and $\vec{k}_\perp$, the wave vector in the plane perpendicular to $\vec{B}$, the ambient dc magnetic field. The absorption of fast waves by ion cyclotron resonance can be deduced from terms being proportional to $E_+ = E_x + iE_y$, $\ell \geq 1$. Carrying out the integration over $\vec{k}$, we readily deduce the following relevant expression for ion cyclotron resonance absorption:

$$\frac{\partial f_o(\vec{v})}{\partial t} = \frac{\pi Z^2 e^2}{8m^2|k_\parallel|} \sum_\ell \left|E_+\right|^2 \frac{1}{v_\perp} \frac{\partial}{\partial v_\perp} v_\perp^2 J_{\ell-1}^2 \left(\frac{k_\perp v_\perp}{\omega_{ci}}\right) \delta\left(v_\parallel - \frac{\omega - \ell\omega_{ci}}{k_\parallel}\right) \frac{1}{v_\perp} \frac{\partial f_o}{\partial v_\perp} \quad (47)$$

where the summation is over ion cyclotron harmonics. Using cylindrical coordinates and azimuthal symmetry, we may readily proceed with the integration in Eq.(45):

$$P = \frac{\pi^2 Z^2 e^2}{8m|k_\parallel|} f_{o\parallel}(v_{\parallel Res})|E_+|^2 \int_0^\infty f_{o\perp} \frac{\partial}{\partial v_\perp} \left[v_\perp^2 J_{\ell-1}^2 \left(\frac{k_\perp v_\perp}{\omega_{ci}}\right)\right] dv_\perp \quad (48)$$

where we integrated by parts twice for convenience. For a Maxwellian, Eq.(48) can be readily integrated for small arguments of the Bessel function and we obtain

$$P = \frac{\omega_{pi}^2 |E_+|^2}{16\pi^{1/2} k_\parallel v_{ti}} b_i^{(\ell-1)} \frac{\ell}{(\ell-1)! 2^{(\ell-1)}} \exp\left[-\frac{(\omega - \ell\omega_{ci})^2}{k_\parallel^2 v_{ti}^2}\right] \quad (49)$$

which is exactly the same result as Eq.(22). Thus, quasi-linear theory gives the same result as Eq.(3) for a Maxwellian distribution. We can now use Eq.(48) to integrate over a Maxwellian distribution to all orders of the finite ion Larmor radius. In particular, taking the derivative in Eq.(48) and integrating by parts, Eq.(48) can be integrated exactly and we obtain

$$P = \frac{\omega_{pi}^2 |E_+|^2 \ell}{16\pi^{1/2} k_\parallel v_{ti}} \left[I_{\ell-1}(b_i) + \frac{b_i}{\ell}\left(I_\ell(b_i) - I_{\ell-1}(b_i)\right)\right] e^{-b_i} \exp\left[-\frac{(\omega - \ell\omega_{ci})^2}{k_\parallel^2 v_{ti}^2}\right] \quad (50)$$

where $I_\ell(b_i)$ is the modified Bessel function of order ℓ and argument b_i. The small Larmor radius limit of Eq.(50) results from the first term of the square bracket (expand $I_{\ell-1}(b_i)$) and it agrees with Eq.(49). It is easy to generalize Eq.(50) to include absorption on an energetic minority ion species (compare Eq.(50) with (41)). These examples show the power of using quasi-linear theory for calculating power absorption to all orders of the

Larmor radius, and for arbitrary distribution functions. In particular, larger values of b_i lead to stronger absorption, and the result is that cyclotron harmonic resonances in the plasma lead to energetic ion tail production due to quasi-linear diffusion.

We can also use Eqs.(45) and (46) to calculate power absorption due to electron Landau damping and electron transit time magnetic pumping. In this case we take the $\ell = 0$ terms and expand the Bessel functions for small arguments (i.e., small electron Larmor radius expansion) before integrating over $v_\perp$. In particular, we use $J_o(k_\perp v_\perp/\omega_{ce}) \simeq 1$, $J_{\pm 1}k_\perp v_\perp/\omega_{ce}) = \pm k_\perp v_\perp/2\omega_{ce}$, and we obtain in Eq.(46)

$$v_\perp^2 |\Theta_o|^2 = \frac{1}{4}\frac{k_\perp^2}{\omega_{ce}^2}|E_y|^2 \left| v_\perp^2 - \frac{2i\omega_{ce}}{k_\perp}\frac{\omega}{k_\parallel}\frac{E_z}{E_y}\right|^2 . \tag{51}$$

For a Maxwellian plasma, we can use the results of Eq.(12) to express the ratio of E_z/E_y in terms of the dielectric constants. Integrating over the energy (Eq. 45) we obtain a result for the power absorbed which is identical to Eq.(5b), and the damping rates are identical to those obtained previously. We can also examine the case of high perpendicular energies and in this case the first term (TTMP) would dominate in Eq.(51). We see that in general, Eq.(51) includes TTMP (the first term) electron Landau damping (the second term) and the cross term (the product of the first and second terms). Equation (51) has been used in Fokker-Planck code calculations in connection with fast wave current drive in anisotropic plasmas which cannot be described by a Maxwellian.[20] In general, using the fast magnetosonic wave it is difficult to distort the distribution function from a Maxwellian for $\omega/k_\parallel v_{\parallel e} \lesssim 2$, which is the usual region of interest for reasonable single pass damping (i.e., several percent per pass)[9]. Thus, for most cases of practical interest the results obtained earlier (Eqs. 13, 21) usually suffice. We shall now discuss some recent experimental results regarding direct electron heating by the fast magnetosonic wave, and point out possible consequences such as fast wave current drive.

XII. Recent Experiments on Direct Electron Heating
by Magnetosonic Waves

Magnetosonic waves have been used for nearly two decades in tokamaks to heat ions, and indirectly electrons, by cyclotron (harmonic) resonance (commonly called "ICRF heating"). The earlier experimental results which emphasize minority heating and cyclotron harmonic heating have been summarized by P.L. Colestock,[21] and more recent references can be found in the AIP Conference Proceedings on RF heating.[22] In this section we wish to mention the very recent results obtained on the DIII-D tokamak on direct electron heating by magnetosonic waves,[23-26] and its potential use for current drive in tokamaks. In addition, we should mention the early experiments in this area on JFT-2M[27] and JET[28,] where a small fraction of the RF power (~20%) was absorbed directly by electrons. In the case of JET, the direct electron heating was limited to less than single pass absorption since a strong ion cyclotron resonance layer was present near the plasma center. In JFT-2M the single pass absorption was very weak owing to the high parallel phase velocity of the injected waves ($\omega/k_\parallel v_{te} \gtrsim 1.7$).

In DIII-D, a four-strap antenna was used to launch fast waves at 60 MHz.[23,25] The magnetic field is varied in the range $B_T = 1$ to 2 Tesla, so that in the deuterium plasma $f/f_{CD} = 8 - 4$ for this range of fields. The $n_\parallel$ spectra of the waves launched peaks at $n_\parallel \simeq \pm 9$ for $(o, \pi, 0, \pi)$ phasing of adjacent antenna elements (with secondary peaks at $n_\parallel = \pm 2.5$) and for $(\pi/2)$ phasing between adjacent current straps the spectrum peaks at $n_\parallel \simeq 5$ with high directionality. The spectral width is typically $\Delta n_\parallel \simeq \pm 2$ about the central value. The evanescent region between the $n_\parallel^2 = R$ cut-off layer and the antenna surface preferentially couples the lower-$n_\parallel$ components while toroidal effects upshift the coupled $n_\parallel$ spectra approximately by the inverse aspect ratio (1/2.7). The Landau absorption condition ($\omega/k_\parallel v_{te} \sim 1$) is satisfied for

$$T_e(keV) \simeq (250/n_\parallel^2)$$

or for $n_\parallel \simeq 9$, $T_{eo} \simeq 3$ keV, and for $n_\parallel \simeq 5$, $T_{eo} \simeq 10$ keV. Thus, for the heating phasing we have ideal absorption conditions, while for the current drive phasing, additional heating (for example, with ECH power) is desirable. It should be recognized, however, that for the heating case the effective $n_\parallel$ spectrum may be lowered by evanescence, while toroidal effects would upshift it back to near its original value. On the other hand, for current drive phasing $n_\parallel \simeq 5$ couples well and toroidal upshifts would result in an effective spectrum of $n_\parallel \sim 7$. Thus, we expect that for $T_{eo} \simeq 5$ keV the absorption of the current drive spectra should be satisfactory.

In Fig. 2 we display heating results with symmetric phasing ($\Delta\phi = \pi$) of the antenna straps. This figure shows good heating of electrons, as well as increase of the stored energy in the plasma. In Fig. 3 the calculated effective energy confinement time, normalized to ITER-P-89 scaling, is shown, as well as the calculated single pass absorption. Note that multiple pass absorption must be effective since nearly all power must be absorbed to account for the observed confinement time. There is no apparent dependence on the magnetic field, which indicates very effective multiple pass absorption. On the other hand, the exponential Landau factor is very important, as may be seen in Fig. 4. There is essentially no heating below a threshold electron temperature, while above this threshold there is rapid increase in the heating effectiveness. Finally, in Fig. 5, evidence of H-mode confinement is indicated by pure electron absorption of the fast magnetosonic wave. The threshold power is comparable to that of neutral injection or ECH power. Thus, the pure electron heating regime with the fast magnetosonic wave has been clearly verified in these seminal experiments. Very recently these experiments have been extended to $(\pi/2)$ phasing, and the existence of fast wave current drive has been demonstrated.[25,26]

XIII. Summary and Conclusions

In this treatise we summarized some aspects of plasma heating by magnetosonic waves in the ion cyclotron range of frequencies. We have considered wave propagation in the simplest possible way, namely that of slab geometry, to estimate the power absorption by electron Landau damping, transmit time magnetic pumping, and cyclotron (harmonic) damping. For simplicity, we have used the local approximation of both the real and the imaginary parts of the wave disperions relationship. It should be pointed out that toroidal wave propagation codes have recently verified these results.[30] We have shown that the simple approach outlined by Stix in 1961, namely calculating the $Re(\bar{J} \cdot \bar{E})$ contribution, gives the same result as quasi-linear theory derived by Kennel and Engelman in 1966. While we have concentrated on the relatively "new" concept of electron absorption of magnetosonic waves (advocated by Stix in his 1975 paper), a brief summary of cyclotron harmonic and minority species absorption was also presented. No attempt was made at mathematical rigor (for this see, for example, the book by Swanson[31]). In fact, the approach taken was that of an "experimentalist," who needs relatively simple absorption formulae which can be used for practical estimates. The limitation of this approach was pointed out wherever applicable, in particular, power "loss" by mode-conversion in the presence of dissipation is hard to estimate without a more rigorous approach. The conditions for efficient single pass absorption were obtained. While in present day experiments mainly minority heating is employed, for future applications the importance of absorption by electron Landau damping must be emphasized. A natural by-product of such absorption is the possibility of noninductive current drive by fast waves. In this context, cyclotron (harmonic) abosrption must be regarded as perhaps and undesirable and "parasitic" loss mechanism since it will reduce the already marginal current drive efficiency. Thus, it is clear that efficient bootstrap current generation must be part of any kind of future steady-state tokamak reactor scenario.

Finally, a very recent experiments on JFT-2M, JET, and in particular on DIII-D have clearly verified the practicality of direct electron absorption of the fast magnetosonic wave in the ICRF regime. In addition, initial results on fast wave current drive have been obtained. In the high temperature reactor regime single pass absorption of fast waves by electrons will be sufficient (of the order ~50% or greater). One of the issues still to be resolved is the competing mechanism of absorption of the fast wave by alpha particles for frequencies $\omega \gtrsim \omega_{c\alpha}$. Clearly, optimizing the choice of wave frequency will be important for good current drive efficiency.

Acknowledgement

This work was supported by the U.S. Department of Energy Contract No. DE-AC02-78ET-51013. The author wishes to thank Dr. C. Petty of General Atomics for his permission to reproduce Figs. 2-5 from Reference 24.

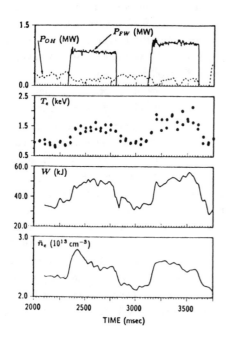

Fig. 2: Time history of (a) net fast-wave and Ohmic power, (b) central electron temperature, (c) plasma stored energy, and (d) line-averaged density. The coupled fast-wave power is 80% of the net power ($I_p = 300$ kA, $B_\tau = 1.6$ T, $\kappa = 1.4$, deuterium discharge with $\approx 2\%$ hydrogen fraction). (After Ref. 24)

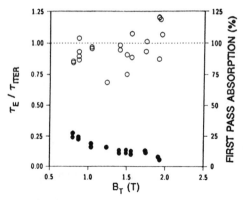

Fig. 3: Energy confinement time (open circles) normalized to the ITER-89 power-law scaling relation as a function of the toroidal magnetic field for high-aspect-ratio plasmas ($R = 1.86$ m, $a = 0.48$ m, $I_p = 300$ kA, $\bar{n} \approx 2.5 \times 10^{19}$m^{-3}, $\kappa = 1.4$). Also shown is the calculated first-pass absorption (solid circles). (After Ref. 24)

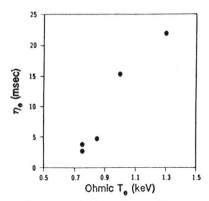

Fig. 4.: Electron heating effectiveness as a function of the target electron temperature [$I_p = 510$ kA, $B_\tau = 1.4$ T, $\bar{n} = (1.1 - 1.9) \times 10^{19}$m^{-3}, $\kappa = 1.9$]. (After Ref. 24)

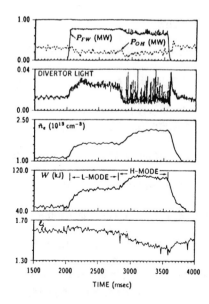

Fig. 5: Time history of (a) net fast-wave and Ohmic power, (b) divertor light, (c) line-averaged electron density, (d) plasma stored energy, and (e) normalized internal inductance. A transition form L-mode confinement to H-mode confinement occurs around 2850 msec, and the H mode continues until the fast-wave power is removed ($I_p = 510$ kA, $B_\tau = 1.0$ T, $\kappa = 1.9$). (After Ref. 24)

References

(1.) T.H. Stix, Nucl. Fusion **15**, 737 (1975).

(2.) N.J. Fisch, Rev. of Modern Physics, **59**, 175 (1987).

(3.) R. Bickerton, J.W. Connor, J.B. Taylor, Nature, **229**, 110 (1971); also, A.A. Galeev, Sov. Phys. JETP **32**, 752 (1971).

(4.) C.F. Karney, F.W. Perkins, and Y.C. Sun, Phys. Rev. Lett. **42**, 162 (1979).

(5.) M. Porkolab, General Atomics DIII-D Physics Memorandum 8813, (March 24, 1988).

(6.) M. Porkolab, General Atomics DIII-D Memorandum 8817 (November 11, 1988).

(7.) S.C. Chiu, V.S. Chan, R.W. Harvey and M. Porkolab, Nucl. Fusion **29**, 2175 (1989).

(8.) M. Porkolab, in Ninth Topical Conf. on Radio Frequency Power in Plasmas (Charleston, SC, 1991) (AIP, NY, 1991) 197.

(9.) D. Moreau, J. Jacquinot and P.P. Lallia, in Contr. Fusion and Plasma Heating (Proc. 13th Eur. Conf., Schliersee, 1986) Vol. 10C, p. 421 (1986).

(10.) M. Porkolab and M. Mayberry, General Atomics Technologies Inc., DIII-D Physics Memorandum D3DPM 8713 (July 15, 1987).

(11.) J. Jacquinot, Plasma Physics and Contr. Fusion, **33**, 1453 (1991).

(12.) T.H. Stix, The Theory of Plasma Waves, McGraw-Hill, New York (1962).

(13.) F.W. Perkins, Nucl. Fusion **17**, 1197 (1977).

(14.) S. Texter, S. Knowlton and M. Porkolab, Nucl. Fusion **26**, 1279 (1986).

(15.) A.K. Ram, A. Bers, V. Fuchs, R.W. Harvey and M.G. McCoy, Radiofrequency Heating and Current Dive in Fusion Devices, Brussels, July 7, 1992, (Proc. Europhysics Topical Conf., Vol. 16E, p. 201, 1992).

(16.) P.E. Moroz, N. Hershkowitz, R.P. Majeski, and J.A. Tataronis, University of Wisconsin Report PLR-92-4 (Feb. 20, 1992). To be published in Phys. Fluids.

(17.) C.F. Kennel and F. Engelmann, Phys. Fluids **9**, 2377 (1966).

(18.) M. Brambilla and M. Ottaviani, , Plasma Phys. and Controlled Fusion **27**, 1 (1985).

(19.) T. Imai, et al., Proc. 13th Int. Conf. on Plasma Phys. and Contr. Nucl. Fusion Research (Washington, 1990) IAEA, Vienna **1**, 645 (1991).

(20.) A. Becoulet, D. Moreau, G. Giruzzi, B. Saoutic, in Fast Wave Current Drive in Reactor Scale Tokamaks, (Proc. IAEA Tech. Comm. Meeting, Arles, France, 1991) p. 62.

(21.) P.L. Colstock, " An Overview of ICRF Experiments," in <u>Wave Heating and Current Drive in Plasmas</u>," Gordon and Breach, New York [Eds. V.L. Granatstein and P.L. Colestock] 1985, p. 55.

(22.) <u>Radio Frequency Power in Plasmas</u>, AIP Conference Proceedings, 244 (Charleston, SC, 1991; D.B. Batchelor, Ed.), Proc. 190 (Irvine, CA, 1989; R. McWilliams, Ed.); Proc. 159 (Kissimmee, FL, 1987; S. Bernabei and R.W. Motley, Eds.).

(23.) C.C. Petty, R.I. Pinsker, M.J. Mayberry et al., in <u>Radio Frequency Power in Plasmas</u>, American Institute of Physics Proceedings 244 [Ed. D.B. Batchelor] p. 96.

(24.) C.C. Petty, R.I. Pinsker, M.J. Mayberry, et al., Phys. Rev. Lett. **69**, 289 (1992).

(25.) R. Prater, R.A. James, C.C. Petty, et al., Preprint of Invited Talk presented at the "Topical Conf. on Radiofrequency Heating and Current Drive in Fusion Devices," Brussels, Belgium, July 8-10, 1992 (Report GA-A21003, to be published).

(26.) R.I. Pinsker, C.C. Petty, M. Porkolab, et al., presented at Fourteenth Int. Conf. on Plasma Physics and Contr. Nucl. Fusion Res., Sept. 30, 1992, Würzburg, Germany, IAEA CN-56/E-2-4.

(27.) T. Yamamoto et al., Phys. Rev. Lett. **63**, 1148 (1989).

(28.) L.G. Eriksson and T. Hellsten, Nucl. Fusion **29**, 875 (1989).

(29.) M.J. Mayberry, S.C. Chiu, M. Porkolab, et al., in <u>Radio-Frequency Power in Plasmas</u>, AIP Conf. Proceedings 190, Roger McWilliams, Ed.) 1989, p. 298.

(30.) See, for instance, P.T. Bonoli and M. Porkolab, in <u>Radio Frequency Power in Plasmas</u> in AIP Conf. Proceedings 244 (D.B. Batchelor, Ed.) 1991, p. 155.

(31.) D.G. Swanson, <u>Plasma Waves</u>, Academic Press, Inc., San Diego, 1989, p. 242-269.

Foundations of ICRF Heating--
A Historical Perspective

Joel C. Hosea
Plasma Physics Laboratory, Princeton University
Princeton, New Jersey 08543

Abstract

Tom Stix has made many major contributions to the development of understanding of a wide array of rf heating and diagnostics methods, both in experiment and theory.   In recognition of his profound influence on ion cyclotron range of frequencies (ICRF) heating research, this paper is focused on two major building blocks contributed by him which served to help guide and quantify the research toward establishing ICRF heating as a viable technique for the reactor regime: 1) the formalism for quantitative evaluation of antenna loading contained in his 1962 text book[1] and 2) his Fokker-Planck analysis for heating of ions and especially minority species ions in his 1975 Nuclear Fusion paper.[2] Importantly, his work from the mid 1950s to the mid 1970s from which these two building blocks derive, provided a solid basis for the rapid developing ion cyclotron heating research in the 1970s and helped to guide that research to definitive demonstration of the viability of the minority ion heating regime as a reactor heating method by the end of the decade.[3]

I. Introduction

Parallel ion cyclotron heating research programs had been under development in the United States[4] and the Soviet Union[5] for several years at the time of declassification in 1958. The programs in both countries were directed primarily toward devising practical means of exciting waves in plasma devices and toward theoretical understanding of the antenna coupling to the waves and of the propagation of the waves in the plasma to the resonant damping zones.   The early work through the decade of the 1960s was directed almost entirely to cylindrical systems and very little power deposition theory was attempted to quantify the damping due to ion resonance at the cyclotron layer, it being assumed to be sufficient to fully absorb the wave energy over the lengths involved in most ongoing experiments. It is during this period that cold plasma wave theory and its applications were presented in two classic texts--The Theory of Plasma Waves by Tom Stix[1] and Waves in Anisotropic Plasma by W. Allis, S. Bucksbaum, and A. Bers.[6] Notably, this theory and its practical application to cylindrical plasma formed the basis of the Model C Stellarator coupling studies which ultimately demonstrated the validity of the theory at relatively low plasma temperature (tens of ev and less). This research not only demonstrated that the coupling could be quantitatively evaluated using the formalism of Stix with the complete cold plasma theory (including finite electron inertia) but also that the so-called perpendicular ion cyclotron resonance[1] results in a mode conversion to an asymptotically electrostatic branch of the plasma dispersion relation in a low temperature plasma which indeed supports propagating waves.[7]

At the end of the 1960s, with the turn of the major fusion research programs toward the toroidal geometry of the tokamak and away from the cylindrical geometry utilized in the racetrack stellarator designs, the coupling and power deposition studies were redirected toward toroidal geometry. A host of tokamaks were employed to develop ion cyclotron heating during the 1970s as indicated in Fig. 1.[8] Some of the significant innovations included the introduction of partial turn coils to provide major radius separation of wave excitation and propagation zones from resonance damping zones,[9] the demonstration of the qualitative applicability of the cold plasma cylindrical coupling theory adapted to toroidal geometry to include toroidal eigenmodes,[9,10] analysis of damping characteristics for fast waves in toroidal geometry,[11] and importantly here the analysis of the effect of minority ions on wave damping and resulting ion species and electron heating in the plasma.[2] It is important to realize that it was generally recognized up to 1975 that second harmonic damping in deuterium plasmas was better than expected (see especially Ref. 12) and that Tom Stix provided a quantitative minority hydrogen ion cyclotron heating explanation of this result in his 1975 paper. This result not only provided the basis for optimizing the minority ion heating regime on the PLT tokamak[3] to demonstrate its viability for reactor application but also provided a basis for extending the Fokker-Planck analysis to include other regimes and effects of fusion products.

II.   Stix Formalism for Quantitative Evaluation of Antenna Loading

Tom Stix introduced the concept of utilizing a periodic induction coil about a cylindrical plasma column to excite electromagnetic waves in the plasma (see Fig. 5.5 of Ref. 1). In the conventional mirror geometry, these waves could then propagate along the plasma column and damp in the lower magnetic field regions in the vicinity of the cyclotron resonance layers as indicated in Fig. 2. In his analysis of this coil geometry, he replaced the discrete coil current elements with a representative coil surface current distribution as illustrated in Fig. 2. He then decomposed this surface current into a Fourier current spectrum to permit analysis of the excitation of the discrete propagating modes supported by the cylindrical plasma geometry and the plasma dispersion relation for the appropriate region of the CMA diagram. For the slow Alfven wave used primarily in mirror (Stellarator) research, $\Omega = \omega/\omega_{ci} < 1$ is appropriate and, as indicated in Fig. 2, corresponds to the 13th region of Stix's CMA diagram (Ref. 1, Fig. 2-1). He employed contour integration (Ref. 1, Fig. 5.6) to arrive at the field amplitudes of the discrete eigenmodes for a given coil current, I. Subsequently, he summed the mode contributions to rf power coupled by integrating $E \cdot J^*$ over the coil volume (J^* is the coil surface current density): $P = -1/2 \text{ Re } \int_v E \cdot J^* dv$. Finally, the loading resistance for the antenna is obtained simply as $R = C P/I^2$ where C is a constant depending on the geometry of the coil feeds.

With this formalism in hand, it was shown early in the Model C Stellarator program that a Faraday shield was required to avoid "anomalous" loading.[13] With the Faraday shield in place, the peak antenna loading at plasma densities above a few times 10^{12} cm^{-3} did follow the theory as formulated by Stix (see Eq. 55 on p. 101 for n=1 in Ref. 1) with electron inertia set to zero (electron background unable to support wave potentials along the magnetic field).[14] However, at lower plasma densities in this study, peak loading dropped off as density was decreased and the peak approached $\Omega = 1$ in contradiction to the Stix analysis.

In order to obtain a quantitative evaluation of the loading predictions of Stix's formalism, a research effort was undertaken on the Model C Stellarator to measure loading very accurately at low rf power and to compare it with the cold plasma theory predictions obtained with the Stix formalism but including finite electron inertia[15] The application of finite electron inertia introduced a second slow electromagnetic root to the dispersion relation[6] (approaching electrostatic at large wavenumber) which coupled to the single root of the Stix analysis to form a mode conversion and establish a slow branch as indicated in Fig. 3. Importantly, this added root provided a mode conversion in the plasma density gradient for a given parallel wavenumber (in place of a perpendicular resonance) and supported the only propagating modes for a low density plasma. Hence, a quantitative comparison of the theory derived from this formalism to the measurements not only should validate the formalism but should also demonstrate the existence of the mode conversion. The results obtained definitively support both affirmations as indicated in Fig. 4.[7]

III. Fokker-Planck Model for Ion Cyclotron Heating in a Multi-Ion Species Plasma

In the tokamak experiments of the 1970s, the emphasis of ion cyclotron heating research shifted to fast wave heating scenarios which utilized propagating waves on the fast wave branch of the dispersion relation (Fig. 3) and which involved damping of waves on the second harmonic resonance of deuterium (Fig. 2, $\Omega = 2$). This shift was based partly on the results that heating tended to be more effective at this resonance[9] and partly on the desire to launch waves from the more accessible outer region of the tokamak.[16]

As mentioned in the Introduction, many advances were made in antenna development and in demonstrating the continued applicability of the cold plasma wave theory to fast wave excitation in toroidal geometry. However, perhaps the most significant development was the quantitative evaluation of rf heating effects on plasma ions and electrons taking into account toroidal and minority ion effects.

From the observations of loading and heating efficiency at the second harmonic of deuterium on the ST[9] and TFR[12] tokamaks (among others in Fig. 1), it was clear that the prevailing theory for second harmonic damping was not complete and that other mechanisms of damping must be considered. Tom Stix's 1975 Nuclear Fusion paper [2] immediately provided the major damping mechanism involved in the experiments-- cyclotron damping by the minority hydrogen ion species contained in the majority deuterium ion plasma. This work coupled with wave propagation and field theory developed by a host of researchers in parallel to show that the fast wave polorization maintained a left hand component at the resonance[17] catapulted the understanding of fast wave heating of toroidal plasmas to a quantitative level in a very short time. Importantly, the minority ion energy distribution was prescribed by Stix directly in terms of measurable quantities assuming all of the absorbed rf power went to the minority ion species which subsequently distributed it according to Fokker-Planck theory and through energetic hydrogen ion losses from the plasma. His energy distribution calculations for a deuterium minority heated at its fundamental in a tritium majority plasma are shown in Fig. 5.

The earlier results on the ST and TFR tokamaks fell into place and a definitive demonstration that the ion tails were indeed formed by the hydrogen minority was first published by Vdovin et al. for the TM-1-VCh tokamak[18] and shown to scale with rf power as prescribed by Stix. The values of ξ deduced were very qualitative estimates due to the large energetic hydrogen ion losses which were present under TM-1-VCh parameters. A quantitative comparison between experiment and theory was obtained later under the much better energetic confinement conditions on the PLT tokamak[19] as shown in Fig. 6.

After being verified experimentally, the heating model presented by Stix was extended to the case of a hydrogen minority in a ^{3}He plasma on the PLT tokamak to demonstrate that the same minority heating physics applied to the tritium majority with deuterium minority case. In addition, although it was not initially possible to measure the ^{3}He tail, successful minority ^{3}He cyclotron heating of a deuterium plasma was also demonstrated on PLT.[20]   This case led to significant energetic 14.7 MeV proton production as shown in Fig. 7[3] which required the presence of a ^{3}He energetic ion distribution consistent with that prescribed by the Fokker-Planck theory.

IV.   Summary

The basis for ICRF wave excitation and heating calculations to which Tom Stix contributed greatly was largely in place at the beginning of the 1980s. His antenna loading formalism has continued to contribute to antenna development studies up to the present time and has permitted engineering attention to be focussed on the crucial issues associated with viable antenna design for survival in the reactor environment. His Fokker-Planck analysis has underpinned the continued optimization of minority heating through the 1980s and into the 1990s: temperatures of 5 keV/4 keV for deuterium ion electrons were obtained in PLT in 1985[21] along with very large ~1 keV sawteeth on the electron temperature; even higher electron temperatures were obtained on JET as early as 1986--7.5 keVwith P_{RF} = 7 MW in conjunction with neutral beam injection at P_{NB} = 2.5 MW--and the large sawteeth on the electrons were stabilized under conditions with a high energetic ion tail population present;[22] and finally, energetic ion populations produced via minority heating have been used to produce toroidal Alfven eigenmodes on the TFTR tokamak[23] to address the steps required to assure that similar effects which might be produced by energetic fusion products do not lead to excessive energy loss from the fusion plasma.

References

1   T. Stix, The Theory of Plasma Waves (McGraw-Hill, New York, 1962).
2   T. Stix, Nuclear Fusion 15 (1975) 737.
3   J. Hosea et al., Proc. 8th Int. Conf. on Plasma Phys. and Contr. Fusion (IAEA, Brussels, 1980) Vol. II, p. 95
4   T. Stix, Phys. Fluids 1 (1958) 308.
5   A. Akheizer et al., Proc. 2nd Int. Conf. on Peaceful Uses of Atomic Energy, (United Nations, Geneva, 1958) 31 p. 99.
6   W. Allis, S. Bucksbaum and A. Bers, Waves in Anisotropic Plasmas (MIT Presss, Cambridge 1963)
7   J. Hosea and R. Sinclair, Phys. Rev. Lett, 23 (1969) 3.
8   J. Hosea, Bull. American Phys. Soc. (1986) 1RM, 1383.
9   J. Adam and the ST Group, Proc. 5th Int. Conf. on Plasma Phys. and Contr. Fusion (IAEA, Tokyo, 1974) Vol. I, p. 65.
10 J. Hosea and W. Hooke, Phys. Rev. Lett. 31, (1973) 150.
11 J. Adam and A. Samain, Fontenay-aux-Roses, Report EUR-CEA-FC-579 (1971) 29.
12 Equipe TFR, Proc. 3rd Int. Meeting on Theoretical and Experimental Aspects of Heating in Toroidal Plasmas (CEA, Grenoble, 1976), Vol. I, p. 87.
13 Rothman et al., J. Nucl. Energy C8 (1966) 241.
14 Rothman et al., Phys. Fluids 12 (1969) 2211.
15 J. Hosea and R. Sinclair, Phys. Fluids 13 (1970) 701.
16 H. Takahaski et al., Journal de Physics 38 (1977) C6-171 (and references therein).
17 F. Perkins, Nucl. Fusion 17 (1977) 1197; D. Swanson et al., Nucl. Fusion 17 (1977) 299; J. Jacquinot et al., Nucl. Fusion 17 (1977) 88; P. Colestock et al., Proc. of 2nd Joint Grenoble-Varenna Int. Sym. (Como, 1980) Vol. I, p. 471; among others
18 V. Vdovin et al.,JETP Lett. 24 (1976) 374.
19 J. Hosea et al., Phys. Rev. Lett; 43 (1979) 1802.
20 J. Hosea et al., Proc. Course on Physics of Plasmas Close to Thermonuclear Conditions (CEA, Varenna, 1979) Vol. II, p. 571.
21 J. Wilson, Proc. 6th Topical Conf. on RF Plasma Heating AIP 129 (1985) 28.
22 JET Team (J. Jacquinot et al.), Proc. 11th Int. Conf. on Plasma Phys. and Contr. Fusion (IAEA, Kyoto, 1986) Vol. I, p. 449.
23 J. Wilson et al., Proc. 14th Int. Conf. on Plasma Phys. and Contr. Fusion (IAEA, Wurzburg, 1992) paper IAEA-CN-56/E-2-2.

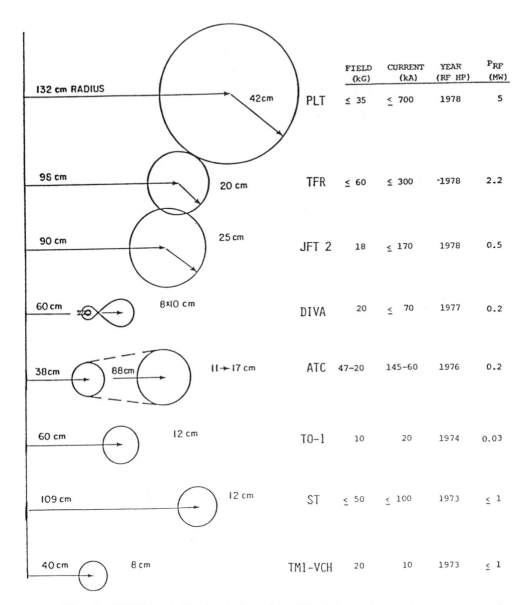

Fig. 1   ICRF heated tokamaks in the 70s.

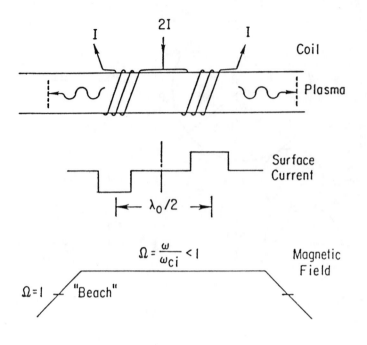

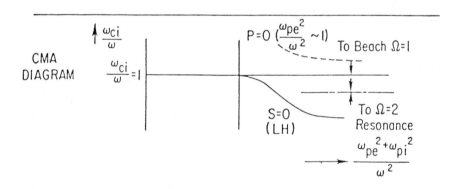

Fig. 2   RF wave considerations in the conventional cylindrical mirror system.

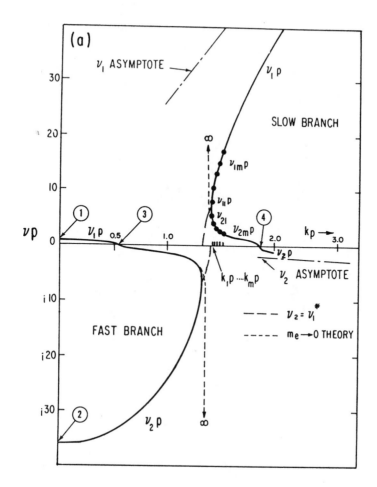

$(n = 10^{13} \text{ cm}^{-3}, \Omega = 0.85)$

Fig. 3 Cold plasma wave theory including effects of electron inertia on the slow wave (Ref. 15, with permission).

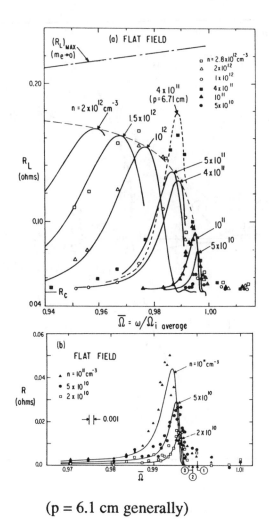

(p = 6.1 cm generally)

Fig. 4   Antenna loading theory compared with measurements for plasma densities between 2×10^{12} cm-3 and 2×10^{10} cm-3. (Ref. 7, with permission). The quantitative agreement demonstrates the presence of the mode converted branch in Fig. 3.

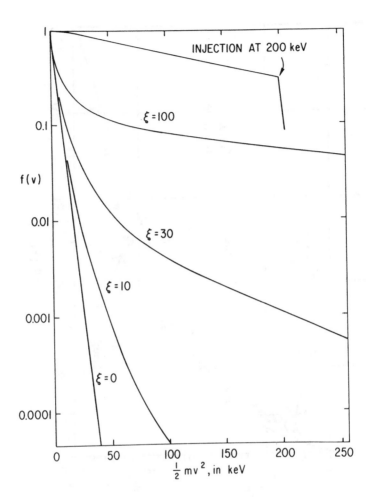

$(\bar{n}_T = 5 \times 10^{13}$ cm^{-3}, $n_D = 2.5 \times 10^{12}$ cm^{-3},
$T = 4$ keV, $Z_{eff} = 3)$

Fig. 5 Fokker-Planck energy distributions for minority deuterium
ion heating in a tritium plasma (Ref. 2, with permission).

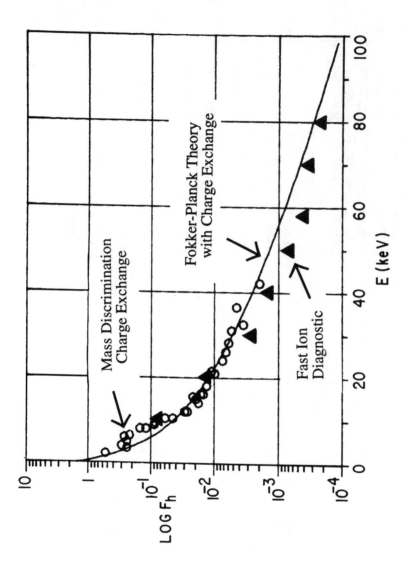

Fig. 6 Hydrogen minority ion distribution in PLT (Ref. 20, with permission).

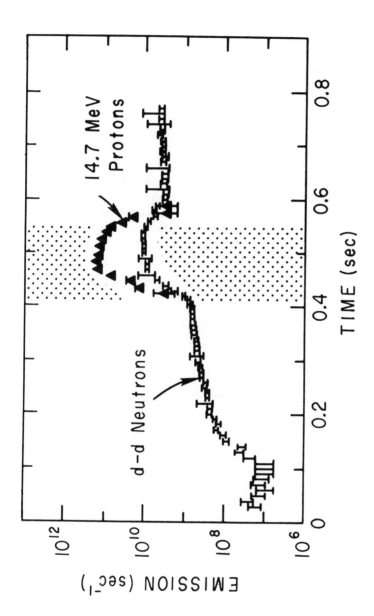

Fig. 7 14.7 MeV proton emission from d (3He, p) α reactions in a deuterium plasma with 3He minority in PLT; P_{rf} = 180 kW, $\bar{n}_e$ = 1.5 × 10^{13} cm^{-3}, n = 3He = 10%. (Ref. 3, with permission)

SCENARIOS FOR THE NONLINEAR EVOLUTION OF BEAM-DRIVEN INSTABILITY WITH A WEAK SOURCE

H.L. Berk and B.N. Breizman
Institute for Fusion Studies, The University of Texas at Austin,
Austin, Texas 78712

ABSTRACT

The problem of a weak source of particles that forms a distribution function that is unstable to a discrete number of modes with the electrostatic bump-on-tail instability taken as a paradigm is considered. Over a wide range of parameters the system produces pulsations, where there are relatively brief bursts of waves separated by longer intervals of quiescent behavior. There are two types of pulsations; benign and explosive. In the benign phase, valid when particle motion is not stochastic, the distribution function is close to that predicted by classical transport theory, and the instability saturates when the wave trapping frequency equals the expected linear growth rate. If the field amplitude reaches the level where orbit stochasticity occurs, the particle diffusion leads to a further conversion of the distribution's free energy to wave energy. This leads to a rapid quasilinear relaxation (a phase space explosion) of the distribution function. Hence the overall response of the system is characterized by a relatively long time interval where the source needs to build up the distribution to its unstable shape as well as provide a sufficient amount of free energy for the instability to grow to the stochastic threshold of particle motion. The particle distribution is then flattened by the quasilinear diffusion in a relatively short time interval to regenerate the cycle.

INTRODUCTION

When energetic particles are present in a plasma their distribution is either readily predictable from classical collisional slowing down and scattering processes or, if instability is sufficiently virulent, the distribution is drastically different from the one that is predicted from classical theory. To assess when anomalous transport arises when there is instability, one has to study a self-consistent nonlinear problem for the evolution of unstable modes simultaneously with the evolution of the fast particle distribution. This generic problem has been a subject of extensive theoretical studies and one of the motivations for developing quasilinear theory[1,2] and weak turbulence theory.[3-5]

This problem has applications to a topic of current interest, the confinement of alpha particles under ignition conditions when Alfvén instabilities are present.[6-12] A steady-state alpha particle distribution forms as a result of the balance between the weak source of high energy particles (produced by the fusion reaction) and the slowing down of these particles by drag with the background plasma. This steady-state distribution has a destabilizing shape that can cause the excitation of Alfvén waves, via the free energy drive of the universal instability. The Alfvén waves that are excited are discrete eigenmodes.

With the motivation of the alpha particle problem, we discuss in this paper the following aspects of the wave-particle interaction: the discreteness of the mode spectrum, damping of the waves from dissipation of the background plasma, the presence of a particle source, and the effects from collisions which bring particles to or take them from the resonant region in a phase space where the particles interact with the waves.

A dominant nonlinear effect in our consideration will be the flattening of the particle distribution function near the resonances. This means that we will neglect the wave-wave nonlinear interactions between the modes and will take into account only the mode interaction with the resonant energetic particles. Another interesting part of the problem is whether one should expect the saturation of the instability at a stationary level or quasiperiodic bursts of waves.

To attempt to understand this problem clearly we will consider a simpler physical problem as a paradigm. We study the bump-on-tail instability, where we include the physical features mentioned above. Namely, the energetic particle distribution is fed continuously, and the modes have discrete phase velocities approximately given by $v_{ph} = \omega_p/k_m$ where ω_p is the plasma frequency and $k_m = 2\pi m/L$ with m an integer and L the plasma length.

SATURATION OF ISOLATED MODES

The first question we address is whether to expect a steady or bursting response when only a single mode is unstable. This problem has been studied in Refs. 10 and 13. In this section, we reproduce the essence of the arguments presented there.

The bump-on-tail instability requires that the slope of the energetic particle distribution, F, be positive in the vicinity of the Cherenkov resonance between the particle and the excited wave, i.e.

$$\frac{\partial F(v)}{\partial v} > 0 \qquad (1)$$

at $kv = \omega$, with k the wave number, ω the wave frequency, and v the energetic particle velocity.

In Ref. 10 a steady-state nonlinear wave was predicted when classical relaxation of energetic particles is accounted for. The solution allows for a balance between the nonlinear particle instability drive and plasma dissipation. As it was shown in Ref. 13, such a solution requires the background damping to be sufficiently weak whereas for stronger background damping rates, the steady-state nonlinear solution is unstable. In this case a new nonlinear scenario emerges. The system no longer maintains a steady-state solution. Instead the response is that of pulsations.

In what follows we will concentrate on the limiting case where the instability manifests itself most clearly. Namely, we assume $\gamma_L \gg (\gamma_d, \nu_{\text{eff}})$, where γ_L is the linear growth rate associated with the distribution function formed from classical relaxation processes in the absence of excitations, γ_d the dissipation rate of the excited wave caused by the background plasma, and ν_{eff} is the rate of reconstruction of the unperturbed distribution function after it has been flattened in phase space by a nonlinear wave. Several mechanisms determine ν_{eff}. Frequently, pitch angle diffusion is the dominant process, and in this case $\nu_{\text{eff}} \approx \nu \omega^2 / \omega_b^2$, where ν is the 90° velocity pitch angle scattering rate, and ω_b the bounce frequency of resonant particles trapped in the wave. If drag determines ν_{eff}, then $\nu_{\text{eff}} \approx \nu_d \omega / \omega_b$ with ν_d the drag rate; while if particle annihilation determines ν_{eff}, then $\nu_{\text{eff}} \approx \nu_a$, with ν_a the particle annihilation rate.

Let us first suppose that $\nu_{\text{eff}} \gg \gamma_d$. In this case the steady-state solution is appropriate. In steady state a wave is found where the power, P, which is transferred from the fast particles to the wave, is given by

$$P \approx \gamma_L \left(\frac{\nu_{\text{eff}}}{\omega_b} \right) WE , \qquad (2)$$

where WE is the energy of the wave. (For electrostatic plasma waves, $WE = \overline{|\delta E|^2}/4\pi$, where the bar refers to time average, δE is the perturbed electric field, and equal energy contributions are taken into account for perturbed electric field energy and perturbed kinetic energy).

Generically, ω_b is proportional to the square root of the wave amplitude and specifically for plasma waves $\omega_b^2 = (e/m)k|\delta E|$. This power is absorbed by background dissipation, $P_d = 2\gamma_d WE$. Hence, with $P - P_d = 0$, the saturated wave amplitude satisfies

$$\omega_b \approx \frac{\gamma_L \nu_{\text{eff}}}{\gamma_d} . \qquad (3)$$

As we assumed $\gamma_d < \nu_{\text{eff}}$, we see that the relaxation process pumps the wave to an amplitude that gives a bounce frequency higher than the linear growth rate.

If $\nu_{\text{eff}} \ll \gamma_d$, the predicted bounce frequency in Eq. (3) is lower than γ_L. In this case the nonlinear steady-state distribution function found in Ref. 10 is unstable, basically to the same linear instability that exists in the unperturbed state. This observation readily follows from closely examining the response of linear theory. The linear growth rate, γ_L, is given by the following expression:

$$\gamma_L = \frac{-2\omega\pi e^2}{|k|m} \, \text{Im} \int dv \, \frac{1}{\omega - kv} \frac{\partial F}{\partial v} . \qquad (4)$$

For a smooth distribution function formed in the absence of nonlinear waves, γ_L reduces to

$$\gamma_L = \frac{2\omega\pi^2 e^2}{|k|m} \int dv \, \frac{\partial F}{\partial v} \delta(\omega - kv) . \qquad (5)$$

In the case $\nu_{\text{eff}} \ll \gamma_d$, the nonlinear distribution function found in Ref. 10 only differs from the unperturbed one in a small resonance region where particles are trapped in the wave. There the distribution is flattened over an area

$$\delta v \approx \omega_b / k \equiv v_b . \tag{6}$$

Outside this region virtually the same F is obtained as in the unperturbed case. Hence, if one attempts to evaluate $\gamma_L(\omega)$ in Eq. (4), with this locally flattened distribution function, one finds that though $\gamma_L(\omega_0) \to 0$ with ω_0 the real frequency of the background oscillation, the value for γ is hardly changed from the value γ_L found in the smooth case (the difference is $\mathcal{O}(\omega_b/\gamma_L)$). Hence the steady-state solution is unstable for sufficiently large γ_d, viz., $\gamma_d \gg \nu_{\text{eff}}$.

This result indicates that the nonlinear response in the $\gamma_d \gg \nu_{\text{eff}}$ limit cannot be a steady state. Instead the following pulsation scenario is envisaged. Suppose the linear bump-on-tail instability with the smooth F distribution develops at the rate γ_L. The distribution function would initially look like the thick solid line in Fig. 1,

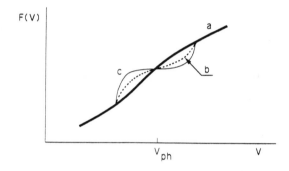

Fig. 1   Time behavior of the bump-on-tail distribution function near a resonant mode's phase velocity. The thick solid curve (a) indicates distribution just prior to relaxation; the thin solid curve (c) is just after relaxation; and the dashed curve (b) is at an intermediate time when the distribution is being reconstituted.

just when instability begins. Then, as basic and straightforward arguments indicate, the wave amplitude will grow until the bounce frequency of the trapped particles reaches the linear growth rate γ_L. The wave flattens the distribution function in the resonant region, which destroys the resonant particle drive in the manner described by O'Neil[14] and Mazitov,[15] and it is depicted by the thin solid curve in Fig. 1. However, with background dissipation present, this wave will now damp according to the equation $dWE/dt = -2\gamma_d WE$. Simultaneously, the classical transport mechanism attempts to reconstitute the unstable distribution function in the flattened region, $\delta v/v \approx \omega_b/\omega \approx \gamma_L/\omega$, at a rate ν_{eff}. Thus the time for the wave energy to disappear is $1/\gamma_d$, while the time for reconstitution is $1/\nu_{\text{eff}}$. After a time $1/\nu_{\text{eff}}$ the distribution is again ready to excite waves to an amplitude where $\omega_b \sim \gamma_L$. During intermediate times $1/\gamma_d < t < 1/\nu_{\text{eff}}$, precursor instability may arise, for example when the distribution is shaped like the dashed curve in Fig. 1. Low amplitude saturation will then occur due to particle trapping with a trapping frequency $\omega_{b1} \approx \gamma_L \nu_{\text{eff}} t < \gamma_L$. However these precursor waves do not destroy the free energy of the distribution in the velocity range

$$\frac{\omega_{b1}}{k} < \left| v - \frac{\omega}{k} \right| < \frac{\gamma_L}{k} \, .$$

Thus, low level precursor waves are expected prior to the largest "crash." After the largest crash, when $\omega_b \approx \gamma_L$, the distribution is again flattened over the interval $\delta v \approx |\delta v_b|$, with $\delta v_b \sim \gamma_L/k$, and then the process described repeats itself with an overall period ν_{eff}^{-1}.

The need for a pulsation scenario can also be explained in terms of energy balance, which shows that it is energetically impossible to sustain a steady excitation level if $\nu_{\text{eff}} < \gamma_d$. Over a long time scale, the average background dissipation can be estimated as $\gamma_d \overline{WE}$, with $\overline{WE}$ the time-averaged wave energy. This dissipation must be balanced by the free energy that is brought to the resonant region by collisions. In a time $1/\nu_{\text{eff}}$ the free energy of the particles is built up and then converted to the maximum wave energy WE_{max} determined from the condition $\omega_b \approx \gamma_L$. This free energy comes from the particle distribution and is equal to the difference in kinetic energy in the distributions (a) and (c) in Fig. 1. Hence the estimate for the feed power into the wave is $\nu_{\text{eff}} WE_{\text{max}}$. Equating the feed power to the average dissipative power gives

$$\overline{WE} = \left(\frac{\nu_{\text{eff}}}{\gamma_d} \right) WE_{\text{max}} \, . \tag{7}$$

Since ν_{eff} is assumed to be much less than γ_d, the average wave energy is much less than the maximum. Such a condition can only be achieved with relaxation oscillations, as depicted in the solid curves in Fig. 2.

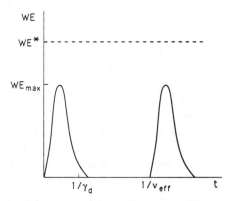

Fig. 2   Pulsating relaxation oscillations. If $\nu_{\text{eff}} < \gamma_d$, pulsations arise as shown by solid curves. If $\nu_{\text{eff}} > \gamma_d$, a steady wave arises with the wave energy saturating at a level $WE^* = (\nu_{\text{eff}}/\gamma_d)WE_{\text{max}}$.

However, as previously discussed,[10] for $\nu_{\text{eff}}/\gamma_d > 1$, the wave energy saturates at a stationary level $WE^* = (\nu_{\text{eff}}/\gamma_d)WE_{\text{max}}$, as depicted by the dashed line of Fig. 2.

MULTIPLE MODES AND PHASE SPACE EXPLOSION

It follows from the previous section that a single unstable mode can only modify the particle distribution function locally. A different picture may arise when there are many unstable modes in the system with the fluctuation level exceeding the stochasticity threshold. This is because particles now really diffuse in phase space and there are no longer barriers to maintain an overall "inverted population" in the vicinity of the resonance region. This regime of the bump-on-tail instability is illustrated in Fig. 3. Below the critical amplitudes for mode overlapping, the situation is depicted in Fig. 3a, where the distribution flattens locally in the shaded regions, with an energy release proportional to $N \lesssim \frac{\omega}{\gamma_L}$, the number of modes. The picture changes drastically, as shown in Fig. 3b, when the resonances overlap. Then all the free energy of the inverted gradient is available to pump the waves to yet higher levels, and to cause strong particle diffusion.

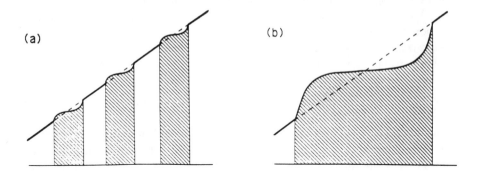

(a)   (b)

Fig. 3 Effect of resonance overlapping. In (a) modes do not overlap, and the relaxed distribution just flattens locally, with the preservation of the general shape of the inverted equilibrium distribution. When modes overlap as in (b), the distribution flattens completely over the entire spectrum, with much larger conversion of free energy to wave energy.

When the amplitudes of excited modes exceed the threshold of resonance overlap, the effect of the waves on the particles is described as quasilinear diffusion. The corresponding diffusion equation for the particle distribution function then has the form:

$$\frac{\partial F}{\partial t} = \frac{\partial}{\partial v} D \frac{\partial F}{\partial v} - \nu_a F + Q(v) \ . \tag{8}$$

Here, the diffusion coefficient, $D(v)$, is related to the spectral density of the wave energy, $W(k)$, by

$$D(v) = \frac{4\pi^2 e^2}{m^2 v} W(\omega_p/v) \,.$$
(9)

The function $W(k)$ is normalized by

$$\int W(k) dk = U$$
(10)

where U is the wave energy per unit volume. The second and third terms on the right-hand side of Eq. (8) describe the source and the annihilation of the fast particles. We choose the source, $Q(v)$, and the annihilation rate, ν_a, to meet the requirement that the "classical" stationary solution of Eq. (8)

$$F = \frac{Q(v)}{\nu_a}$$
(11)

has a sufficiently large positive derivative $\partial F/\partial v$ to drive the bump-on-tail instability in the presence of a background damping. In order to simulate the feature that all potentially unstable modes are in a certain interval of phase velocities ranging from $v_{\min}$ to $v_{\max}$, we set $D = 0$ outside this interval.

After discussing Eq. (8) we will also consider a modified version of this equation in which the annihilation term is replaced by the collisional slowing down term and the source is localized outside the area of quasilinear diffusion (at a velocity v_0 larger than $v_{\max}$)

$$\frac{\partial F}{\partial t} = \frac{\partial}{\partial v} D \frac{\partial F}{\partial v} + \frac{\partial}{\partial v} (\nu v F) + Q(v) \,.$$
(12)

In this equation, the effective collision frequency ν will be taken to decrease with increasing v faster than $1/v$, to provide an instability of the "classical" stationary solution corresponding to a constant particle flux in the velocity space q, so that $Q(v) = q\delta(v - v_0)$:

$$F = \frac{q}{\nu v} \,.$$
(13)

To Eq. (8), we add the equation for evolution of wave energy,

$$\frac{\partial W(k)}{\partial t} = 2\gamma W(k) - 2\gamma_d W(k)$$
(14)

where the first term with

$$\gamma = \frac{2\pi^2 e^2 \omega_p}{k^2 m} \frac{\partial F}{\partial v} \, (v = \omega_p/k)$$

describes the wave excitation by energetic particles, while the second term takes into account background damping. The damping rate, γ_d, is assumed to be much less than the typical linear growth rate produced by the unperturbed stationary distribution (11) at the interval $(v_{\min}, v_{\max})$. Thus the distribution (11) is strongly unstable. The stable stationary solution of Eqs. (8) and (14) differs from (11) due to quasilinear diffusion and is determined from the following equations:

$$\frac{2\pi^2 e^2 v^2}{m\omega_p}\frac{\partial F}{\partial v} - \gamma_d = 0 \tag{15}$$

$$\frac{\partial}{\partial v} D \frac{\partial F}{\partial v} - \nu_a F + Q(v) = 0 . \tag{16}$$

By integrating Eq. (15), $F(v)$ is obtained to within a constant. This constant is found by integrating Eq. (16) from $v_{\min} - \varepsilon$ to $v_{\max} + \varepsilon$ with the boundary conditions $D(v)\frac{\partial F}{\partial v}|_{v=v_{\max}+\varepsilon} = D(v)\frac{\partial F}{\partial v}|_{v=v_{\min}-\varepsilon} = 0$, and we obtain

$$\int_{v_{\min}}^{v_{\max}} \nu_a F \, dv = \int_{v_{\min}}^{v_{\max}} Q \, dv . \tag{17}$$

With this condition taken into account, the solution for $F(v)$ is

$$F = F_1 + F_2 \tag{18}$$

where

$$F_1 = \frac{\displaystyle\int_{v_{\min}}^{v_{\max}} Q \, dv}{\displaystyle\int_{v_{\min}}^{v_{\max}} \nu_a \, dv} \tag{19}$$

$$F_2 = \frac{m\omega_p}{2\pi^2 e^2} \frac{\displaystyle\int_{v_{\min}}^{v} \gamma_d \frac{dv}{v^2} \int_{v_{\min}}^{v_{\max}} \nu_a \, dv - \int_{v_{\min}}^{v_{\max}} \nu_a \, dv \int_{v_{\min}}^{v} \gamma_d \frac{dv_1}{v_1^2}}{\displaystyle\int_{v_{\min}}^{v_{\max}} \nu_a \, dv} . \tag{20}$$

The ratio of F_1 to F_2 is roughly of the order of γ_L/γ_d where γ_L is the linear growth rate for the unstable "classical" distribution function (11). When γ_L is assumed to be much larger than γ_d, one can neglect the velocity dependent contribution to F and then the distribution is nearly constant,

$$F = \frac{\displaystyle\int_{v_{\min}}^{v_{\max}} Q \, dv}{\displaystyle\int_{v_{\min}}^{v_{\max}} \nu_a \, dv} . \tag{21}$$

We now combine Eqs. (15), (16), and (9) to find

$$W(\omega_p/v) = \frac{mv^3}{2\omega_p \gamma_d} \int_{v_{\min}}^{v} (\nu_a F - Q) dv . \tag{22}$$

For the simplified distribution (21) we obtain

$$W(\omega_p/v) = \frac{mv^3}{2\omega_p\,\gamma_d}\int_{v_{\max}}^{v}\left(\nu_a\,\frac{\displaystyle\int_{v_{\min}}^{v_{\max}} Q\,dv_1}{\displaystyle\int_{v_{\min}}^{v_{\max}} \nu_a\,dv_1} - Q\right)dv\;. \tag{23}$$

This equation shows that in quasilinear theory the wave energy density W scales linearly with Q. However, when the source is very weak, the wave energy is insufficient to provide mode overlapping and Eq. (23) is not applicable. In this case, each unstable mode forms a separate island in the phase space and quasilinear diffusion really does not arise, since island-to-island transitions are strongly suppressed.

Let us then study more carefully the cases when most of the time there is no mode overlap. Let E_i be the electric field amplitude of the i-th discrete mode. Then the energy density of this single mode, $E_i^2/8\pi$, can be estimated as

$$\frac{E_i^2}{8\pi} \sim \frac{\omega_p}{vN}\,W \tag{24}$$

where N is the total number of modes. To overlap the neighboring resonances one needs

$$\frac{\Delta v}{v} > \frac{1}{N} \tag{25}$$

where Δv is the velocity perturbation of the particle that resonates with the i-th mode. For Δv we have

$$\Delta v = \sqrt{\frac{evE_i}{m\omega_p}}\;. \tag{26}$$

By combining Eqs. (24)–(26) we find the following criterion of resonance overlapping:

$$W > \omega_p\,\frac{m^2\,v^2}{8\pi\,e^2\,N^3}\;. \tag{27}$$

Taking into account Eq. (23) we rewrite Eq. (27) as a restriction on the particle source

$$Q > \frac{\gamma_d\,m\omega_p^2}{4\pi\,e^2\,N^3 v}\;. \tag{28}$$

We then conclude that quasilinear stationary solution (21), (22) breaks when the intensity of the source is below the critical value given by Eq. (28). The "classical" stationary solution (11) is also inappropriate since we have chosen it to be strongly unstable. This indicates again that the system does not reach a stationary state but rather creates bursts which explosively release the free energy built up by the particle source.

In order to estimate the energy of a burst, we first neglect the particle source and the wave damping. As long as the excited discrete modes do not overlap (Fig. 3a) each of them saturates when the bounce frequency of a resonant particle trapped by the mode reaches the linear growth rate γ. In this regime, one has

$$\frac{\Delta v}{v} = \frac{\gamma}{\omega_p} . \tag{29}$$

As time progresses the source causes the slope of the distribution to build up so that γ increases and $\Delta v/v$ eventually reaches the value $1/N$. At this critical value of γ, the total free energy of the unstable distribution becomes available for the burst (Fig. 3b). This energy can be estimated as the energy that is released through global flattening of the distribution with $\gamma = \gamma_{\text{crit}} \equiv \frac{\omega_p}{N}$:

$$U_{\text{burst}} \sim \frac{m^2 \omega_{pe}^2}{24\pi^2 \, e^2 \, N} \, (v_{\text{max}} - v_{\text{min}})^3 \, \frac{1}{v_{\text{max}} + v_{\text{min}}} . \tag{30}$$

This consideration shows that a weak source is unable to build up a particle distribution with a free energy exceeding the value given by Eq. (30).

By comparing Eqs. (27) and (30) we may note that the wave energy required for mode overlapping is much less than U_{burst}:

$$U_{\text{overlap}} \sim \frac{U_{\text{burst}}}{N^2} . \tag{31}$$

Therefore, the burst is well described by quasilinear theory. This theory predicts complete flattening of the particle distribution within the time of the inverse critical growth rate ω_p/N. During a relatively short time the wave energy builds up to the level given by Eq. (30). Then the waves damp at a rate γ_d, with the distribution function remaining flat since the source is too weak to change the particle distribution within the damping timescale. The third, longer phase, is building up the free energy required for the next burst. The time interval, τ_{rst}, of the restoration, is determined by the energy balance. Hence, τ_{rst} is inversely proportional to the intensity of the source:

$$\tau_{\text{rst}} \sim \frac{U_{\text{burst}}}{mQv_{\text{max}}^3} . \tag{32}$$

It is interesting to note that when $\gamma_{\text{crit}} \ll \gamma_L$ the average power transfer from the particles to the waves is rather insensitive to whether the system reaches quasilinear stationary state or creates bursts. This result is straightforward to observe when one writes the average power transfer to the waves, $\overline{P}_w$, as a difference between the power supplied from the source and the dissipation from annihilation:

$$\overline{P}_w = \frac{1}{T} \int_0^T dt \int_{v_{\text{min}}}^{v_{\text{max}}} \frac{mv^2}{2} \, (Q - \nu_a F) dv , \tag{33}$$

where the averaging period, T, is over many burst periods. This expression only depends on the particle distribution function which, when $\gamma_{\text{crit}} \ll \gamma_L$, is close to plateau (19) in both cases. The bursts of the wave energy are obviously easier to observe than the corresponding small deviations of the particle distribution function from the plateau. It should also be noted that, most of the time between bursts, the distribution function is metastable as shown in Fig. 4.

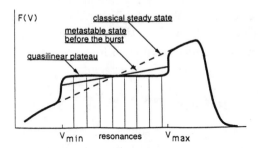

Fig. 4   Metastable distribution formed between the bursts in weak source and particle annihilation model. Burst of instability forms quasilinear plateau. Prior to the burst the distribution is metastable. In this state wave energy is too low to cause mode overlap, and quasilinear diffusion does not occur. When critical slope is reached the distribution relaxes to plateau. The distribution never reaches its classical steady state profile shown by the dashed line.

Therefore, if an appropriate triggering mechanism is available, the system bursts before the accumulated free energy reaches the critical value given by Eq. (30). Such a mechanism occurs naturally in the "slowing down" model described by Eq. (12), as we will now show.

Proceeding to the analysis of the "slowing down" model, we note that, as the "classical" solution (13) of Eq. (12) is assumed to be strongly unstable, the constant flux solution of this equation must incorporate the quasilinear flux:

$$D \frac{\partial F}{\partial v} + \nu v \, F = q \, . \tag{34}$$

This equation must be solved together with Eq. (15), and the solution must satisfy a boundary condition that $F(v)$ is a continuous function at $v = v_{\min}$ since otherwise the flux would be singular at $v = v_{\min}$. Then we obtain

$$F = \frac{q}{\nu(v_{\min})v_{\min}} + \frac{m\omega_p}{2\pi^2 \, e^2} \int_{v_{\min}}^{v} \gamma_d \frac{dv}{v^2} \tag{35}$$

$$D = \frac{2\pi^2 \, e^2 \, v^2 \, q}{m\gamma_d \, \omega_p} \left(1 - \frac{\nu v}{\nu(v_{\min})v_{\min}} \right) - \frac{\nu v^3}{\gamma_d} \int_{v_{\min}}^{v} \gamma_d \frac{dv}{v^2} \, . \tag{36}$$

Using the same arguments as for the annihilation model, we can neglect the last terms in F and D when the factor γ_d/γ_L is small. Thus, the solutions (35) and (36) simplify to

$$F = \frac{q}{\nu(v_{\min})v_{\min}} \tag{37}$$

$$D = \frac{2\pi^2\, e^2\, v^2\, q}{m\gamma_d\, \omega_p}\left(1 - \frac{\nu v}{\nu(v_{\min})v_{\min}}\right) \tag{38}$$

and we note again that the distribution function is almost flat in the range $v_{\min} < v < v_{\max}$ due to quasilinear diffusion.

Using arguments similar to those used to obtain Eq. (28), we find that the quasilinear solution breaks down when

$$q < q_{\text{crit}} \equiv \gamma_d\, \frac{m\omega_p}{4\pi e^2\, N^3} \ . \tag{39}$$

In this regime, the wave energy comes out in bursts. However, the typical energy of a single burst differs from that given by Eq. (30). There is a clear trend that the bursts are initiated near $v = v_{\max}$ where the time averaged distribution function shown in Fig. 5 has a discontinuity.

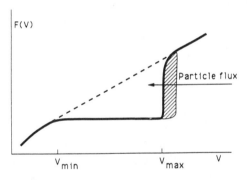

Fig. 5 Stationary distribution function in the "slowing down" model (solid line). "Classical" stationary distribution is shown by the dashed line. When the particle flux is very small, the discontinuity in the distribution function at $v = v_{\max}$ triggers bursts of the wave energy. The shaded area corresponds to the particles which initialize the bursts. The density of these particles is given by Eq. (40).

Though the distribution function shown in Fig. 5 always has positive slopes outside the region $v_{\min} < v < v_{\max}$, it does not excite an instability during a time interval between the bursts since our model limits the phase velocities of all unstable modes to the region $v_{\min} < v < v_{\max}$. However, the instability eventually starts as the discontinuity in the distribution shifts to lower velocities due to collisions.

The number of particles n^* involved in a single burst can be estimated from the condition of the nonlinear saturation of the mode at the upper edge of the spectrum which interacts with this discontinuity and periodically flattens the distribution function near $v = v_{\mathrm{max}}$. This estimate gives

$$n^*/n \approx \left(\frac{\gamma_L}{\omega_p}\right)^{3/2} \tag{40}$$

where n is the plasma density. The burst spreads the particles down to $v = v_{\mathrm{min}}$ releasing the energy of the order of

$$\frac{n^* \, mv_{\mathrm{max}}^2}{2} . \tag{41}$$

The excited waves then damp at a rate γ_d, and the instability "waits" until collisions bring a new portion of particles close enough to $v = v_{\mathrm{max}}$ to create a new burst. As before, when $\gamma_{\mathrm{crit}} \ll \gamma_L$, the average power transfer to the waves, $\overline{P}_w$, is insensitive to whether one has bursts or a truly stationary quasilinear regime. For $\overline{P}_w$ we have

$$\overline{P}_w = q \, \frac{mv_{\mathrm{max}}^2 - mv_{\mathrm{min}}^2}{2} - q \int_{v_{\mathrm{min}}}^{v_{\mathrm{max}}} mv^2 \, \frac{\nu}{\nu(v_{\mathrm{min}})v_{\mathrm{min}}} \, dv . \tag{42}$$

It should be noted that the discontinuity which triggers the bursts only exists because the particle source is located outside the region where waves resonate with particles. A source located inside this region should not cause direct triggering. In this case a larger free energy, up to the estimate of the annihilation model, can be accumulated between bursts.

CONCLUSIONS

We have considered a system where the distribution of energetic particles is formed from the balance of a weak source and a weak relaxation mechanism. The resulting steady-state distribution is assumed to have a shape which tries to destabilize a discrete spectrum of waves. In the absence of energetic particles these waves are supported by the background plasma and are weakly damped. This is a generic problem for many physical cases, and in this paper we discussed in detail the relatively simple case where the waves are electrostatic plasma waves and the beam source forms a bump-on-tail instability.

The critical question in this class of problems is whether the stored energy of the beam is close to the stored energy predicted by the transport properties in the absence of waves. Another interesting question is whether one sees a steady noise level or a pulsating response. Several scenarios have been described where different noise patterns and stored energy are obtained.

In one case we obtain a "benign" scenario where there can be bursting, but the stored energetic particle energy is nearly the same as in the case without excited fields. Then the discrete modes do not cause stochasticity and the distribution builds up to essentially the level predicted by instability free transport theory. If the source strength is large enough stationary steady waves are established. The energy that is being fed to the background plasma through wave dissipation is coming from the particle source. The main alteration of the energetic particle distribution function is in the resonance region. Though the distribution is flattened there, a flux of energetic particles flow through this region because of classical transport processes.

However, if the source strength is too weak, the source cannot maintain a steady-state wave, because at perturbed field levels required for saturation too much energy would be drained by dissipation to the background plasma. In this case "benign" bursts arise that only flatten the distribution function locally but the waves cannot tap the overall free energy source of the energetic particles. The waves only grow up to a natural saturation level where the bounce frequency of particles in the wave equals the linear growth rate. At this stage, the local free energy drive is saturated, and no further energy can be extracted from energetic particles by the wave. Subsequently, the wave damps due to background dissipation, and a time interval determined by classical transport processes needs to elapse before waves re-excite. As the resonant particles cannot move beyond the island boundary of the wave, the overall global distribution function is still close to the one predicted from the simplest transport theory.

If at the estimated level of wave saturation, where the particle bounce frequency equals the growth rate, the resonances of neighboring modes overlap, an entirely different scenario is established. Then particles really diffuse in phase space as described by quasilinear theory. When instability occurs, the particle distribution rapidly flattens to a new constant value over the region of phase space that resonates with the allowable discrete modes. For a sufficiently strong source the noise level can be steady as predicted by quasilinear theory. However, for a weaker source the system is quiescent most of the time. The particle distribution builds up from its flattened state until a certain point is reached where the saturated modes are about to overlap. During this period there can be precursors, but they just lead to the benign saturation previously described, with the overall increase in the distribution function continuing as if there were no oscillations. However, near the point of criticality often determined by mode overlap, the distribution will "explode," and again relax to the flattened quasilinear state, where the cycle repeats. Sometimes the critical point can be determined by other trigger mechanisms.

The characteristic growth and damping rates are: γ_d the damping rate of the background plasma, γ_L the growth rate of the instability as predicted from classical transport, and γ_c the growth rate when overlap occurs. For simplicity we have assumed $\gamma_d \ll \gamma_L$. If $\gamma_c \ll \gamma_L$ we showed that the average properties of the particle distribution is the same for the case when steady-state quasilinear theory is applicable or when the bursting scenario applies. However, if γ_c/γ_L is a finite number (less than unity) there can be considerable difference in the average properties of the bursting scenario compared to the prediction of quasilinear theory.

We define the overall free energy, WF, as the difference between the energy stored in a steady-state distribution and the stored energy of the flattened distribution (the latter energy is essentially the energy predicted to be stored in the quasi-linear calculations). If mode overlap is the trigger mechanism, then the average free energy that can be stored in the bursting scenario is roughly $WF\gamma_c/2\gamma_L$. If $\gamma_c/\gamma_L \ll 1$, very little free energy can be stored. However if $\gamma_c \approx \gamma_L$ the system can store an energy comparable to the classical prediction, just before the explosion flattens the distribution function. Between explosions the system builds up to the critical state.

The application of this picture to more complicated problems, such as Alfvén instabilities, is clear. An analysis for the parameters of this problem has been discussed elsewhere.[16] In the Alfvén problem, the classical transport mechanisms involve drag and pitch angle scattering in velocity space, whereas the quasilinear relaxation primarily involves spatial diffusion. The phase space explosions then imply rapid radial diffusion, which can lead to direct and rapid energetic particle loss to the plasma edge. Such an interpretation is quite compatible with experimental observations.[17,18] Specific predictions as to how alpha particles evolve for a given case will require determining the detailed instability growth rates for the mode spectrum as well as analyzing the mechanisms for particle resonance. Work is in progress to develop quantitative models for the alpha particle confinement when Alfvén instabilities exist.

ACKNOWLEDGMENTS

This work was supported by the U.S. Department of Energy under contract No. DE-FG05-80ET-53088.

REFERENCES

1. A.A. Vedenov, E.P. Velikhov, and R.Z. Sagdeev, Nucl. Fusion Suppl. **2**, 465 (1962).

2. W.E. Drummond and D. Pines, Nucl. Fusion Suppl. **3**, 1049 (1962).

3. A.A. Vedenov, in *Reviews of Plasma Physics*, Consultants Bureau, New York (1967) Vol. 3, M.A. Leontovich (ed.).

4. B.B. Kadomtsev, *Plasma Turbulence*, Academic Press, New York (1965).

5. A.A. Galeev and R.Z. Sagdeev in *Basic Plasma Physics*, North Holland, Amsterdam (1963) Vol. 1, M.N. Rosenbluth and R.Z. Sagdeev (eds.).

6. G.Y. Fu and J.W. Van Dam, Phys. Fluids B **1**, 1949 (1989).

7. C.Z. Cheng *et al.*, in *Proceedings of the Thirteenth International Conference on Plasma Physics and Controlled Nuclear Fusion Research*, International Atomic Energy Agency, Vienna (1991), Vol. 2, p. 209.

8. M.N. Rosenbluth, H.L. Berk, J.W. Van Dam, and D.M. Lindberg, Phys. Rev. Lett. **68**, 596 (1992).

9. F. Zonca and L. Chen, Phys. Rev. Lett. **68**, 592 (1992).

10. H.L. Berk and B.N. Breizman, Phys. Fluids B **2**, 2226 (1990); **2**, 2235 (1990); **2**, 2246 (1990).

11. D.J. Sigmar, C.T. Hsu, R. White, and C.Z. Cheng, Phys. Fluids B **4**, 1506 (1992).

12. H. Biglari and P.H. Diamond, Phys. Fluids B **4**, 3009 (1992).

13. H.L. Berk, B.N. Breizman, and H. Ye, Phys. Rev. Lett. **68**, 3563 (1992).

14. T.M. O'Neil, Phys. Fluids **8**, 2255 (1968).

15. R.K. Mazitov, Zh. Prikl. Mekh. i Techn. Fiz. **1**, 27 (1965).

16. H.L. Berk, B.N. Breizman, and H. Ye, submitted to Phys. Fluids B (IFSR #566).

17. K.L. Wong, Phys. Rev. Lett. **66**, 1874 (1991).

18. W.W. Heidbrink, E.J. Strait, E. Doyle, and R. Snider, Nucl. Fusion **31**, 1635 (1991).

INVESTIGATION OF ELECTROSTATIC WAVES IN THE ION CYCLOTRON RANGE OF FREQUENCIES IN L-4 AND ACT-1

Masayuki Ono

Princeton Plasma Physics Laboratory, Princeton University,
P.O. Box 451, Princeton, New Jersey, 08543, USA

ABSTRACT

Electrostatic waves in the ion cyclotron range of frequencies (ICRF) were studied in the Princeton L-4 and ACT-1 (the Advanced Concept Torus-1) devices for approximately ten years, from 1975 to 1985. The investigation began in the L-4 linear device, looking for the parametric excitation of electrostatic ion cyclotron waves in multi-ion-species plasmas. In addition, this investigation verified multi-ion-species effects on the electrostatic ion cyclotron wave dispersion relation including the ion-ion hybrid resonance. Finite-Larmor-radius modification of the wave dispersion relation was also observed, even for ion temperatures of $T_i \approx 1/40$ eV ($\sim$ the room temperature!). Taking advantage of the relatively high field and long device length of L-4, the existence of the cold electrostatic ion cyclotron wave (CES ICW) was verified. This branch occurs below the ion cyclotron frequency and, in a two-ion-species plasma, shows a resonance at the ion-ion hybrid frequency. With the arrival of the ACT-1 toroidal device, finite-Larmor-radius (FLR) waves were studied in a relatively collisionless warm-ion hydrogen plasma. Detailed investigations of ion Bernstein waves (IBW) included the verification of mode-transformation in their launching, their wave propagation characteristics, their absorption, and the resulting ion heating. This basic physics activity played a crucial role in developing a new reactor heating concept termed ion Bernstein wave heating (IBWH). Experimental research in the lower hybrid frequency range confirmed the existence of FLR effects near the lower hybrid resonance, predicted by Stix in 1965. In a neon plasma with a carefully placed phased wave exciter, the neutralized ion Bernstein wave was observed for the first time. Using a fastwave ICRF antenna, two parasitic excitation processes for IBW -- parametric instability and density-gradient-driven excitation -- were also discovered. In the concluding section of this paper, a possible application of externally launched electrostatic waves is suggested for helium ash removal from fusion reactor plasmas.

INTRODUCTION

Historically, research concerning waves in the ion cyclotron range of frequencies (ICRF) has focused on the electromagnetic waves, such as the ion cyclotron and fast magnetosonic waves, due to their application to plasma heating.[1] Electrostatic waves have been studied primarily to understand phenomena in the plasma edge region during high power rf heating. For example, electrostatic waves can be excited inadvertently through parametric instabilities and parasitic processes associated with high power ICRF heating. These electrostatic waves are apt to heat the plasma edge and produce an influx of impurities. Also, it was pointed out in 1975 that fast waves can undergo mode-conversion into ion Bernstein waves (IBW) near the heating layer, a process that may play a significant role in ICRF power absorption.[2] Earlier, in 1965, Stix proposed that the FLR effects could cause mode-conversion of the electron plasma wave into hot plasma modes near the lower hybrid resonance.[3] Pioneering experiments performed in this area include those on ion acoustic waves,[4] electrostatic ion cyclotron waves,[5] and pure ion Bernstein waves,[6] to name a few. However, there were many more electrostatic modes and processes that needed experimental verification. In particular, data on multi-ion-species effects were largely absent. For example, the existence of ion-ion hybrid resonance phenomena was only inferred through external measurements such as antenna loading.[7] In the area of finite-Larmor-radius modes, the experimental data were essentially limited to the Q-machine experiments. How these electrostatic waves (through linear and/or nonlinear processes) are excited by external antennas was not yet well understood in this frequency range. In the Princeton L-4 device, the study of electrostatic waves in the ion cyclotron range of frequency began in 1975 with an emphasis on parametric instability processes in multi-ion-species plasmas.[8] This work led to the observation of new parametric instability processes in the ion cyclotron range of frequencies.[9-11] At the same time, the research also produced useful information on FLR-modified electrostatic waves in multi-ion species plasmas.[9-11]

After completion of the parametric instability study, the L-4 device was found uniquely suited to verify the existence of the cold electrostatic ion cyclotron wave.[12] Then, in 1979, to examine new concepts for rf current drive[13] and for finite-ion-temperature waves,[14] the L-4 linear device was converted into the ACT-1 torus.[15] Although it is outside of the scope of the present paper, it should be noted that the initiation of current-drive research on ACT-1 has led to a very productive series of experiments starting with the first verification of the lower hybrid current drive in a toroidal plasma.[16] This work evolved into experiments on other types of current drive including fast-wave current drive,[17] helicity-injection current drive in CDX (the Current Drive Experiment performed in the ACT-1 facility),[18] and the pressure-driven-currents in CDX-U (the Current Drive Experiment - Upgrade facility replaced the ACT-1 facility in 1990).[19] Going back to the investigation of electrostatic waves in ACT-1, using warm hydrogen plasmas, the physics of ion Bernstein waves was investigated extensively.[20-24] Emerging from this study came a promising reactor heating concept utilizing externally launched IBW.[14,25] Later experiments in ACT-1 then found long elusive plasma modes including the hot plasma waves (HPW) excited near the lower hybrid mode-conversion layer[24,26] and the well-known neutralized ion Bernstein wave.[27,28] Parasitic IBW excitation processes relevant for the ICRF heating were also examined.[29,31]

This chapter intends to review the topic of electrostatic waves in the ion cyclotron range of frequencies, to summarize research work performed in L-4 and ACT-1 in this area, and to guide interested readers to the original manuscripts for more detail, many of which are Ph.D. theses. In the present paper, we first review the pertinent dispersion relations. Then mode conversion and mode transformation process (applicable to electrostatic waves) are discussed. Experimental parameters important for the L-4 and ACT-1 experiments are described. We next present some experimental highlights, starting with the laboratory determination of the dispersion relation for various waves, mode-transformation and mode-conversion processes, IBW heating experiments, and ICRF-relevant parasitic IBW excitation processes. Finally, we suggest using externally launched IBW and/or CES ICW to remove helium ash from fusion-reactor plasmas.

ELECTROSTATIC WAVE DISPERSION RELATION

The quasi-electrostatic wave dispersion relation in the ion cyclotron range of frequency including the finite-Larmor-radius effect can be written as,[1]

$$n_\perp^2 K_{xx} + (n_\parallel^2 - K_{xx})K_{zz} = 0 . \qquad (1)$$

where K_{xx}, and K_{zz} are elements of the plasma dielectric tensor which in a non-drifting Maxwellian plasma can be expressed as

$$K_{xx} = 1 + \frac{\omega_{pe}^2}{\Omega_e^2} + \sum_i \frac{\omega_{pi}^2}{\omega - k_{\parallel} V_i} \frac{e^{-b_i}}{b_i} \sum_{n=1}^{\infty} n^2 I_n (Z_n + Z_{-n}) \qquad (2)$$

$$K_{zz} = 1 + \frac{2\omega_{pe}^2}{k_{\parallel}^2 V_e^2} (1 + \xi_{e0} Z_0) \qquad (3)$$

where the summation i is over ion species, n is the harmonic number, Z_n is the plasma dispersion function[32] of argument $\xi_{\sigma n} \equiv (\omega - n\,\Omega_\sigma) / k_{\parallel} V_\sigma$, σ denotes species, $V_\sigma \equiv (T_\sigma / m_\sigma)^{1/2}$ and I_n is the modified Bessel function with argument $b_i \equiv k_{\perp}^2 T_i / m_i \Omega_i^2$. Here $n_{\perp} \equiv ck_{\perp}/\omega$ and $n_{\parallel} \equiv ck_{\parallel}/\omega$. The finite-Larmor-radius (FLR) wave formalism was established in 1960's by Bernstein.[33] The generalized dispersion relation was later given in Stix.[1]

To illuminate the multi-ion species and FLR effects, it is instructive to expand terms in Eq.(2) for small $b_i \equiv (1/2)<k_{\perp}\rho_i>^2$, a particularly good approximation for $\omega \approx 2\Omega_i$. Retaining the leading order terms in b_i and assuming large ξ_{in}, one obtains a quadratic equation in $k_{\perp}^2$ as

$$a k_{\perp}^4 + b k_{\perp}^2 - c = 0 \qquad (4)$$

where

$$a = \sum_i 3 \frac{\omega_{pi}^2 (T_i / m_i)}{(4\Omega_i^2 - \omega^2)(\omega^2 - \Omega_i^2)}$$

$$b = 1 - \sum_i \frac{\omega_{pi}^2}{(\omega^2 - \Omega_i^2)}$$

$$c = k_{\parallel}^2 \frac{\omega_{pe}^2}{\omega^2} \qquad \text{for } \frac{\omega}{k_{\parallel}} > V_{Te}, \text{ or}$$

$$= -2 \frac{\omega_{pe}^2}{V_{Te}^2}. \qquad \text{for } \frac{\omega}{k_{\parallel}} < V_{Te}$$

The coefficients a, b, and c represent the finite-Larmor-radius (ion thermal) correction term, the cold ion term, and the cold (or hot) electron term, respectively. For multi-ion-species plasma, we assume species 1 to have the highest ion cyclotron frequency. In an axisymmetric system such as tokamak, one can often assume $n_{\parallel}$ to be a constant or only slowly varying. We shall therefore examine here the wave dispersion relation with the $k_\perp$ as a variable. The solution for Eq. (4) is

$$k_\perp^2 = \frac{-b \pm \sqrt{b^2 + 4ac}}{2a} \tag{5}$$

We shall now consider various limiting cases of Eq. (5).

A. Waves in cold electron plasmas ($\omega / k_\parallel > V_{Te}$).

For cold electrons, c is positive. There are then four main propagating modes depending on the signs of b and a. Examining these four limiting cases:

1. Electron plasma wave or EPW ($\omega > \omega_{pi}$)

In a sufficiently low density plasma ($\omega > \omega_{pi}$), b is positive and a is small. Then, $k_\perp^2 \approx c/b$ or

$$k_\perp^2 \left(1 - \sum_i \frac{\omega_{pi}^2}{(\omega^2 - \Omega_i^2)} \right) - k_\parallel^2 \left(\frac{\omega_{pe}^2}{\omega^2} \right) = 0, \tag{6}$$

which is the well-known electrostatic lower-hybrid-wave dispersion relation. The wave exhibits a resonance ($k_\perp \to \infty$) at the lower hybrid frequency, $\omega \approx \omega_{pi}$. For intermediate densities such that $\omega_{pe}^2 >> \omega^2 >> \omega_{pi}^2, \Omega_i^2$, the dispersion relation becomes particularly simple, $\omega \approx (k_\parallel / k_\perp)\omega_{pe}$. It then follows that $\partial\omega/\partial k_\perp \approx - \omega/k_\perp$; the wave is a backward propagating wave (the wave phase velocity points in the direction opposite to the wave group velocity). It also exhibits "resonance cone" behavior[34,35] [the direction of the group velocity is the same for all wave numbers, i.e., $v_{g\perp} / v_{g\parallel} = (\partial\omega/\partial k_\perp)/(\partial\omega/\partial k_\parallel) = - \omega /\omega_{pe}$]. Measurement of the resonance-cone angle, $\theta \equiv \tan^{-1} [v_{g\perp}/v_{g\parallel}]$, which depends only on the electron density for a given wave, often gives a convenient way of determining the plasma density in a laboratory plasma. The understanding of the launching and propagation of electron plasma waves is highly relevant to lower-hybrid wave heating, and detailed investigations[36,37] have been carried out for that reason.

2. Ion Bernstein wave ($\omega \approx$ few Ω_i and $\omega < \omega_{pi}$)

For $\Omega_i < \omega < 2\Omega_i$, a is positive, and one can see a continuous evolution of propagating solutions ($k_\perp^2 > 0$) from the low-density electron plasma wave (EPW) regime where $b \approx 1$ (i.e., $k_\perp^2 \approx c/b \approx k_\parallel^2 \omega_{pe}^2/ \omega^2$) to the higher density ($\omega < \omega_{pi}$) ion Bernstein wave (IBW) regime for $b < 0$. In the limit of $b << 0$, one obtains $k_\perp^2 \approx$ |b |/a, or for a single ion species plasma,

$$k_\perp^2 \approx (4\Omega_i^2 - \omega^2) / 3(T_i / m_i) \tag{7}$$

where the electron term drops out entirely and the dispersion relation becomes independent of plasma density. The IBW dispersion relation exhibits a cut-off behavior $(k_\perp \to 0)$ as $\omega \to 2\Omega_i$. It is also clear that the perpendicular IBW phase and group velocities are proportional to the ion thermal velocity, V_{Ti}. And again, the wave is a backward propagating wave. For a two ion species plasma, the IBW exists between the ion-ion hybrid frequency, ω_{ih}, defined as $\omega_{ih}^2 = (\omega_{p1}^2 \Omega_2^2 + \omega_{p2}^2 \Omega_1^2) / (\omega_{p1}^2 + \omega_{p2}^2)$ and the lower ion cyclotron frequency, Ω_2. The propagating bands are therefore $\omega > \Omega_1$ and $\omega_{ih} > \omega > \Omega_2$ Solving Eq. (1) in full, one obtains the IBW propagating branches for various harmonic frequencies. See, for example the paper by Swanson.[38] In the limit of $n_\parallel \to 0$, the electron term drops from the dispersion relation altogether, leaving the "pure" ion Bernstein wave where the wave oscillation is sustained by the ion-Larmor-radius dynamics alone. In this limit the wave dispersion relation is simply $K_{xx} = 0$. J. Schmitt was able to observe this wave in a cesium plasma by using a wire exciter which was strung carefully along a magnetic field line in order to satisfy the $n_\parallel \approx 0$ condition.[6] Typically, the IBW shows a cut-off behavior as ω approaches $n\Omega_i$ from below. Similarly, a resonant behavior appears when $n\Omega_i$ is approached from above.

3. Hot plasma waves (HPW) in the lower hybrid wave frequency range ($\omega \gg \Omega_i$ and $\omega > \omega_{pi}$)

It turns out that Eq. (5) can also describe[3] FLR effects for the lower-hybrid wave ($\omega \gg \Omega_i$). The term $a \approx -3\omega_{pi}^2 (T_i / m_i) / \omega^4$ in Eq. (5) becomes negative. The only propagating solutions are when $b \approx 1 - \omega_{pi}^2 / \omega^2$ is positive or $\omega_{pi}^2 < \omega^2$. One solution of Eq. (5) is the electron plasma wave or the lower hybrid wave. The other root describes the dispersion relation of the hot plasma wave, $k_\perp^2 \approx b / |a|$ or

$$k_\perp^2 = \frac{\omega^4 - \omega_{pi}^2 \omega^2}{3 \omega_{pi}^2 (T_i / m_i)} \tag{8}$$

It is easy to show that this mode is a forward propagating mode. It should be noted that this mode can also exists in the hot electron plasma since the dispersion does not depends on the electron term.

4. Cold electrostatic ion cyclotron wave, or CES ICW ($\omega < \Omega_i$ and $\omega < \omega_{pi}$)

In this regime, since $\omega < \Omega_i$, b is positive, and the solution is $k_\perp^2 \approx c/b$. The form is the same as for the EPW. The dispersion relation can then be written as

$$k_\perp^2 \approx \frac{k_\parallel^2 (\omega_{pe}^2 / \omega^2)}{\sum_i \omega_{pi}^2 / (\Omega_i^2 - \omega^2)} \tag{9}$$

The CES ICW dispersion relation shows a cut-off behavior ($k_\perp \to 0$) as $\omega \to \Omega_i$. In a two-ion-species plasma, the wave exhibits resonance behavior ($k_\perp \to \infty$) at the ion-ion hybrid frequency[7], ω_{ih}, where the denominator of Eq. (9) vanishes. The propagating bands are $\omega < \Omega_2$ and $\omega_{ih} < \omega < \Omega_1$. One should point out that the propagation bands of the CES ICW are complementary to those of the IBW. The CES ICW is backward propagating and exhibits resonance-cone behavior.[34,35] One can show from Eq. (9) that the cone angle $\theta \equiv \tan^{-1}[v_{g\perp}/v_{g\|}] \approx \Sigma_i \, [\omega_{pi}/\omega_{pe}](\Omega_i^2/\omega^2 - 1)^{-1/2}$ depends only on the values of local ω/Ω_i and ion concentration Since ω/Ω_i is an accurately known quantity, measurement of the cone angle in this instance is able to provide a convenient method for determining the ion concentration.[12] Since this mode can be excited in about the same frequency band as the ion cyclotron wave (the slow wave), it may be related to the infamous Mode X observed during early ion cyclotron heating experiments carried out in the C-Stellarator with no Faraday shielding for the antenna.[39] That this mode can be easily excited by an external antenna was, in fact, demonstrated in L-4.[12]

Further analysis including the electromagnetic terms shows that CES ICW cannot penetrate beyond the Alfvén resonance. Thus it is not suitable for core plasma heating. On the other hand, it will be pointed out in the concluding section that this mode may be well suited for edge heating, or may be used for the removal of helium ash or impurity ions.

B. Waves for hot electron plasmas ($\omega / k_\| < V_{Te}$.).

For hot electrons, c is negative and there are two well-known propagating roots in the negative b region ($\omega > \Omega_i$ and $\omega < \omega_{pi}$). Equation. (5) can then be written as

$$k_\perp^2 = \frac{|b| \pm \sqrt{b^2 - 4a|c|}}{2a} \qquad (10)$$

1. Electrostatic ion cyclotron wave ($\omega > \Omega_i$)

The negative root, $k_\perp^2 \approx |c|/|b|$ or

$$k_\perp^2 \approx \frac{(\omega_{pe}^2 / V_{Te}^2)}{\sum\limits_i \omega_{pi}^2 / (\omega^2 - \Omega_i^2) - 1} \qquad (11)$$

which is the electrostatic ion cyclotron wave (ES ICW).[5] This mode is insensitive to the plasma density and the phase velocity is proportional to the ion acoustic velocity, $C_s \approx (T_e / m_i)^{1/2}$. The dispersion relation asymptotically approaches that of the ion acoustic wave for $\omega \gg \Omega_i$ since the wave can be considered unmagnetized. The ES ICW dispersion relation exhibits a cut-off behavior ($k_\perp \to 0$) as $\omega \to \Omega_i$.[5] In a two-ion-species plasma, the wave exhibits a resonance behavior ($k_\perp \to \infty$) as it approaches the ion-ion hybrid frequency, ω_{ih}, and the lower hybrid frequency, $\approx \omega_{pi}$, from below.[9]

The propagating bands are $\Omega_2 < \omega < \omega_{ih}$ and $\omega > \Omega_1$. As can be shown from Eq. (11), the ES ICW is a forward-propagating mode.

2. Neutralized ion Bernstein wave ($\omega > \Omega_i$)

The positive root, $k_\perp^2 \approx |b| / a \approx (4 \, \Omega_i^2 - \omega^2) / 3(T_i / m_i)$ has the same dispersion relation as the ion Bernstein wave, Eq (7). However, in this case, the electrons are hot. Because their rapid movement along the field line neutralizes the plasma, this mode is termed the neutralized ion Bernstein wave. The propagating bands of the neutralized IBW are similar to those of the ES ICW. As an IBW, it is a backward-propagating mode.[27,28]

C.   Mode-conversion and mode-transformation processes

At the point where the square root term of Eq.(5), ($b^2 + 4ac$), becomes zero, the two propagating modes [the + root and - root of Eq (5)] merge. This coming together of two different types of waves [i.e., one forward (fw) and one backward (bw)] in the propagating side of the conversion point (i.e., $b^2 + 4ac > 0$) has been termed mode conversion.[3] It represents a true wave singularity in that the group velocity goes to zero at this point. As the wave numbers must match at the conversion point, $k_\perp^2 \approx -b/2a$, the wave group velocity must cross zero. Recall that the group velocities of those two waves start with opposite signs (i.e., they comprise a fw and a bw). Because of this singular behavior, the conversion efficiency is usually not 100%. The wave could be partially reflected. In some situations, the wave could tunnel to the other side if the thickness of the evanescent layer is sufficiently small. This type of problem has been studied in detail for ICRF heating.[2,40] There are also many examples of mode conversion among electrostatic waves: EPW (bw) into HPW (fw); further mode conversion of HPW (fw) into IBW (bw); and ES ICW (fw) into neutralized IBW (bw). For lower hybrid waves, the EPW can mode convert to a HPW, and the converted HPW can again mode convert to an IBW which can then be thermalized by the plasma. This possibility of double-mode-conversion for the lower hybrid heating was first pointed out by Stix.[3]

Mode-transformation on the other hand describes a process where a plasma wave, as it propagates in a non-uniform plasma, undergoes an intact transformation into another type of wave due to a change in the plasma parameters (e.g. density or magnetic field gradient).[14,20] Mode transformation occurs when waves of similar type (two forward waves or two backward waves) are joined together. As long as the WKB condition is satisfied, the mode transformation efficiency is 100%. There is no wave singularity for this case. It is easy to verify this point since the wave number stays finite and the group velocity always retains the same sign (and stays finite). The mode stays on the same root of Eq. (5) but changes its property due to the change in the plasma parameters. Three cases of the mode transformation process among the backward waves are discussed here: 1. EPW into IBW, when b changes sign from positive to negative at the lower hybrid resonance, $\omega \approx \omega_{pi}$; 2. EPW into CES ICW has a rather fuzzy transition at $\omega \approx \omega_{pi}$ since b is always positive here; and 3. CES ICW into IBW, when b changes sign from positive to negative at $\omega \approx \omega_{ih}$. Although not observed experimentally (to the author's knowledge), there appears to be a mode-transformation process that connects two forward waves for $\omega \gg \Omega_i$ (negative a), ES ICW and HPW at b=0. By utilizing this process, one might excite a desired wave with good efficiency by actually exciting another type of wave perhaps launched more easily by an external antenna. For IBW heating, one typically uses the EPW $\rightarrow$ IBW mode-transformation process, transforming a launched electron plasma wave into the ion Bernstein wave for which one expects better wave accessibility to the hot-dense plasma core. One can also launch CES ICW via the EPW$\rightarrow$CES ICW mode-transformation.

A comparison of mode transformation and mode conversion is shown in Fig. 1 where the wavenumber is plotted as a function of the plasma position. Density increases going away from the antenna. The lower curve shows the mode transformation of an EPW into an IBW for $\Omega_i < \omega < 2\Omega_i$ where a is positive. One sees a continuous evolution of the refractive index ($k_\perp^2 > 0$) from the low density electron plasma wave (EPW) regime, where $b \approx 1$ (i.e., $k_\perp^2 \approx c/b$), to the higher density ($\omega < \omega_{pi}$) ion Bernstein wave (IBW) regime where $b \ll 0$ (i.e., $k_\perp^2 > |b|/a$). The cold plasma resonance disappears for a finite ion temperature plasma. By keeping the higher order terms, one can show that mode transformation also occurs for the waves launched at higher harmonics, provided that the ion temperature in the transformation region, near $\omega \approx \omega_{pi}$, is sufficiently high. In

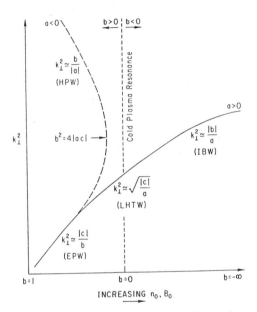

Fig. 1. Behavior of wave dispersion relation near the lower hybrid resonance layer (b = 0). The solid curve shows a mode-transformation process for $\omega < 2\Omega_i$ (a > 0). The dashed curve shows a mode-conversion process typical for $\omega \gg \Omega_i$ (a < 0).

general, for the n-th harmonic launching case, the higher $k_\perp \rho_i$ values require that T_i for a smooth transformation go up as $10^{(n/2-1)}$ in the transformation region.[21]

The dashed upper curve in Fig. 1 shows the mode-conversion process for $\omega \gg \Omega_i$ with a < 0 relevant for lower hybrid heating. One can see that EPW branch connects to HPW mode at the mode-conversion point. It should be noted that near the turning point, the WKB treatment breaks down and a proper treatment of the differential wave equation is required.[3]

III. Experimental setup

A. Experimental parameters of L-4 and ACT-1

Electrostatic waves in the ion cyclotron range of frequency were studied in the L-4 linear device (1975-79) and the ACT devices (1980-1985). A brief description of the relevant plasma parameters for L-4 and ACT-1 follows.

L-4 was a steady-state linear device with a uniform magnetic field extending over approximately 230 cm. The axial magnetic field strength could rise to 4.2 kG and spatial field ripple was below 0.5 %.[11] As shown in Fig. 2, plasma was produced by a hot tungsten filament plasma source[41] located at one end of the device. Emitted electrons are mirror confined by a large number of small permanent magnets placed with alternating polarity on the wall of the filament source chamber. The emitted electrons produce plasma by ionizing the neutral background gas. However, the ionization efficiency was relatively small (since the required ionization path length was usually much longer than the device length) and the resulting plasma was only weakly ionized ($\approx 1\%$). Typical plasma parameters used in the experiments are as follows: plasma density, $n_0 \leq 3 \times 10^{10} cm^{-3}$; electron temperature, $T_e \approx 2 - 5$ eV; ion temperature, $T_i \leq 1/10$ eV; plasma drift velocity $\leq 0.1\ c_s$; and the plasma density fluctuation, $\delta n/n \approx 1 - 2\%$ in the plasma center and 2 - 5 % at the edge of the plasma.

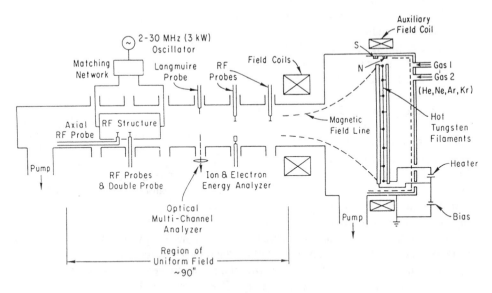

Fig. 2. Schematic of experimental set-up in the L-4 linear device.

ACT-1 (Advanced Concept Torus-1) was a simple toroidal machine with major and minor radii of 59 cm and ~ 6 cm, respectively, and a plasma circumference of about 350 cm.[15] A schematic drawing is given in Fig. 3. One of the unique features of ACT-I was the excellent plasma access provided by the twenty-six 10×40 cm large rectangular windows. The magnetic field ripple was < 1% in the central region of the plasma and the steady-state toroidal magnetic field was typically used at 5 kG. Plasma was created by a hot tungsten filament strung in the high field region as shown in Fig. 4.[17] Emitted electrons circulate around the torus many times before being lost to the wall. The relatively long electron path length (3.7 meters times the number of transits around the torus can often exceed 100 m) insures efficient ionization of the background neutral gases. As shown by Fig. 4, the plasma consists of a narrow vertical

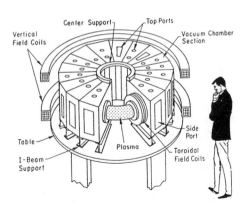

Fig. 3 Schematic of the ACT-1 device.

band with hot electrons near the filament while a diffuse lower-density plasma filled the remaining space bounded by the conducting limiter. As the diffuse plasma is free of hot electrons and is nearly Maxwellian, most of the wave propagation experiments were conducted in this region. The ∇B-drift-generated vertical currents were allowed to

flow to the conducting plasma boundaries (or limiters), and in this way the equilibrium of the plasma was maintained. This type of equilibrium produced plasmas that were relatively uniform in the vertical direction. The measured radial plasma drift was small, < 0.1 $c_s \approx 10^5$ cm/sec. Density fluctuations were typically 5 - 10%. A good stable discharge could be maintained even for gauge neutral pressure as low as 7×10^{-6} Torr, resulting in a warm-ion ($T_i \leq 2$ eV), $T_e \approx$ few eV, highly ionized ($\leq 30\%$), low collisionality hydrogen plasma with $n_e \leq 10^{11}$cm^{-3}.

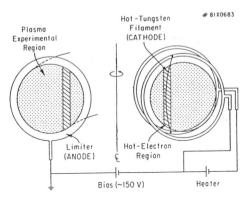

Fig. 4. Schematic of the tungsten filament plasma source in ACT-1.

In the ion cyclotron range of frequencies, it is often important to use a strong magnetic field to attain a high wave frequency (noting that $\omega \sim \Omega_i$) and avoid excessive wave dissipation processes such as ion collisions. Similarly, for cold electron modes it is important to avoid electron Landau damping. In a typical laboratory plasma, it is not always possible to investigate *cold-electron* ion cyclotron modes since the excitable wave parallel phase velocity is not large enough to avoid electron Landau damping, $\omega / k_{\parallel} = f \lambda_{\parallel} > 2 - 3 \ V_{Te}$. If the plasma temperature is a few eV ($V_{Te} \approx 1 \times 10^8$ cm/sec), the wave parallel phase velocity must therefore exceed $2 - 3 \times 10^8$ cm/sec. If one assumes that it is possible to excite a wave with a well defined $\lambda_{\parallel} \approx 50$ cm (this value must be small compared to the plasma size), then the wave frequency must be at least 4

- 6 MHz. This means that the ion cyclotron frequency must be of the order of a few MHz. For hydrogen, the value is reasonable, B = 2 - 3 kG. However, the required value goes up linearly with ion mass. If one were to use a cesium plasma ($m_i \approx 40$) as in Q-machines, the required field becomes almost prohibitively large (even though T_e is low $\approx 0.1 - 0.2$ eV). It was for this reason that a careful alignment of the wire exciter was essential for the pure IBW excitation experiment in the Q-machine.[6] Generally, it was advantageous to use low-mass ion plasmas such as hydrogen and helium for the investigation of cold-electron ion cyclotron waves. Fortunately, both the L-4 and ACT-1 devices had a relatively high (steady-state) magnetic field of 4 - 5 kG. For the investigation of hot electron modes, the requirement is much less stringent. It is however important to keep the parallel wave phase velocity near the valley of ion and electron Landau damping, $V_{Ti} \ll \omega / k_\parallel \ll V_{Te}$. In this case, it was often advantageous to go to heavier ion plasmas such as neon or argon since it is easier to establish the low Landau damping condition.

B.   Ion concentration measurements for the multi-ion species plasmas.

A plasma can naturally contain more than one ion species, e.g. the impurity ions. It is also well known that a hydrogen discharge plasma usually contains three types of ions, H_1^+, H_2^+, and H_3^+. For the investigation of ion cyclotron waves, it is important to determine the concentration of respective ion species, particularly when examining multi-ion-species phenomena such as the ion-ion hybrid resonance. In L-4, using a neon-helium plasma, a spectroscopic method was used to measure the ion concentrations.[11] Figure 5 shows the measured wavelength of an ES ICW in a helium-neon mixture plasma, plotted as a function of the measured concentration. A strong ion concentration dependence is seen. The solid curve, which presents the calculated value from the wave dispersion relation, shows a good agreement with experiment.

It was later discovered that the cold electrostatic ion cyclotron wave (CES ICW) is particularly well suited for the ion concentration measurements [see Eq. (9)].[12] In ACT-1, in addition to the CES ICW method, the ion concentration of hydrogen was also deduced by analyzing the different branches of the IBW dispersion relation.[22]

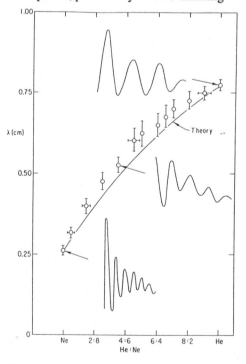

Fig. 5. Ion concentration measurement in the L-4 helium-neon plasma.

C. Warm ion plasmas in ACT-1

In order to investigate finite ion temperature effects, one clearly needs a plasma with a finite ion temperature. But due to the presence of the neutral particles, the iontemperature of the weakly ionized filament discharge plasma in L-4 was quite low, $T_i < 0 1$ eV. In such plasmas, the ionization fraction is relatively small – often well below 1 %.   The neutrals, continuously in contact with the vacuum vessel wall (at room temperature), are quite effective in cooling the ions.   The electron temperature on the other hand is relatively high, in the range of few eV. Electrons remain hot since the electron-ion collisional energy transfer rate is much lower than the corresponding ion-ion or ion-neutral rates. Indeed, low ion temperature behavior was observed in the dispersion relation of the electrostatic ion cyclotron waves in the L-4 device. As shown in Fig. 6, the measured dispersion relation (circles) agrees well with the theoretical value (dashed curves) for an ion temperature of 1/40 eV, i.e., room temperature.[10,11] But even at this low ion temperature, it is interesting to note that the ES ICW dispersion relation shows a clear FLR gap near $\omega \approx 2\Omega_i$.

The new type of plasma obtained in the ACT-1 toroidal device proved well suited to the full exploration of FLR waves. As described above, the combination of high ionization and low neutral-cooling rates yielded warm ions in the ACT-1 plasmas with ion temperatures approaching the electron temperature (a few eV). Indeed, it was found that the ion temperatures scaled roughly inversely with the neutral fill pressures.[21,22,24] In Fig. 7, the

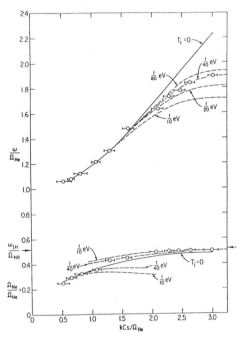

Fig. 6. Electrostatic ion cyclotron waves in L-4.

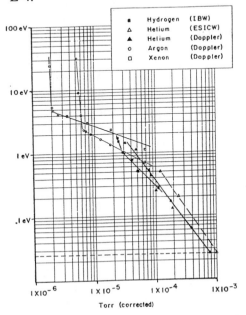

Fig. 7. Summary plot of ion temperatures as a function of absolute fill pressure for various operating gases in ACT-1.

measured ion temperatures for plasmas with various ion species are plotted versus the absolute neutral pressure. Doppler measurements of ion lines as well as the FLR dispersion relations of the ES ICW and IBW were used for the ion temperature measurements. As shown in the figure, by adjusting the neutral fill pressure, it was possible to vary the ion temperature over almost two orders of magnitude in the experiment, providing a method of control that was highly useful in the investigation of FLR effects.

INVESTIGATION OF WAVE DISPERSION RELATIONS

In this section, we summarize the investigation of electrostatic waves in the ion cyclotron range of frequencies in L-4 and ACT-1. Due to the space limitation, we only highlight the results which can be considered the experimental "first" and refer the readers to the original papers and theses manuscripts for more detail.

A. Electrostatic Ion Cyclotron Waves

Electrostatic ion cyclotron waves (ES ICW) in a two-ion-species plasma were investigated in the L-4 device.[9-11] The experiments confirmed that in a two-ion-species plasma, the relative motion of the ions of the two species is able to drive a new type of parametric instability.[10] The process had been predicted by theory.[8] A byproduct of this investigation was the verification of the ES ICW dispersion relation in a two-ion-species plasma. In particular, as seen in Fig. 6, the ion-ion hybrid resonance together with the FLR modification of the dispersion relation near the second harmonic frequency were both observed, even though $T_i \approx$ 1/40 eV.

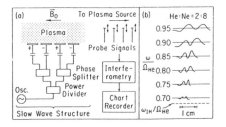

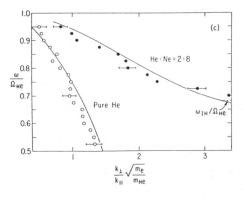

B. Cold Electrostatic Ion Cyclotron Waves

Utilizing the unique capability of the L-4 device (long device length, high field, and relatively cold electrons), the CES ICW could be excited by a slow-wave antenna in a helium-neon plasma. The experimental set-up is shown in Fig 8 (a). The slow-wave antenna

Fig. 8. Investigation of cold electrostatic ion cyclotron waves in L-4. (a) Simplified schematic of wave propagation setup. (b) Radial interferometer output for various values of ω / Ω_{He} (as labeled) in He:Ne = 2:8 plasma. (c) Wave dispersion relation for two ion concentrations (as labeled). Dots are experimentally measured values and solid curves are theoretical values.

consisted of four alternately phased elements comprising two axial wavelengths with $\lambda_{\parallel} = 62$ cm. As noted above, this cold mode can be launched quite easily from an external antenna (provided that it satisfies the cold electron condition) through the EPW $\rightarrow$ CES ICW mode-transformation process (at $\omega \approx \omega_{pi}$). The propagation characteristics, the wavelengths, and the damping were measured by a set of radial probes. Figure 8(b) displays a typical radial interferogram. From such data, the radial wavelength can be seen to decrease rapidly as the wave frequency approaches the ion-ion hybrid frequency, $\omega \rightarrow \omega_{ih} \approx 0.68\ \Omega_{He}$, confirming resonance behavior. The measured dispersion relation is shown in Fig. 8(c) for two concentrations (as marked). The solid curves present theoretically computed values that include finite electron temperature effects. The agreement between theory and experiment is excellent. Backward propagation and resonance cone behavior were both confirmed. Also, near the ion-ion hybrid frequency, enhanced wave damping was measured. This same mode was later utilized in ACT-1 for measuring the hydrogen ion concentration, since the propagation angle depends only on the ion concentration (for a given value of magnetic field and wave frequency).[21]

C. Ion Bernstein Wave ($\omega / k_{\parallel} > V_{Te}$)

The ACT-1 toroidal device was used for a detailed experimental determination of the IBW dispersion relation, Eq.(1). In Fig. 9(a), the measured wave interferogram is shown as a function of the normalized frequency, ω/Ω_H. The wave exhibits a cut-off behavior ($\lambda \rightarrow \infty$) for $\omega \rightarrow 2\Omega_H$. In Fig. 9(b), the wave interferometer output is shown as a function of phase shift. The wave phase front moves toward the antenna while the wave packet is moving away from the antenna, confirming the backward-propagating character of the IBW. The experimental dispersion relation is plotted in Fig. 9(c). The solid curves show the theoretical values, Eq. (1), for various ion temperatures. Best agreement is seen for $T_i = 1.5$ eV. Such measurements can therefore be used to assess the hydrogen bulk-ion temperature for a plasma.[22] In Fig. 10, a more complete IBW dispersion relation measurement is shown. The CO_2-scattering data

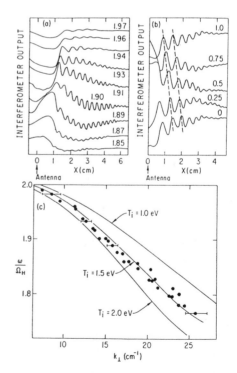

Fig. 9. Ion Bernstein wave identification in ACT-1.

CO_2-scattering data points, shown by the circles, agree well with the probe data (solid triangles).[22,24] The multiple branches of the dispersion relation reflect the presence of deuterium-like and tritium-like hydrogen molecular ions, H_2^+, and H_3^+. Corresponding theoretical dispersion curves are also shown. Fig. 15. Observation of lower-hybrid mode-converted hot-plasma modes in ACT-1.

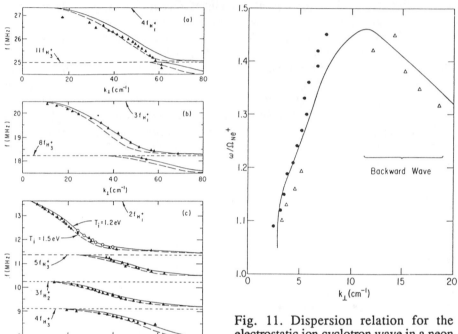

Fig. 10.   A complete IBW dispersion relation in ACT-1 hydrogen plasma.

Fig. 11. Dispersion relation for the electrostatic ion cyclotron wave in a neon plasma in ACT-1.

D. Neutralized Ion Bernstein Wave ($\omega / k_{\parallel} < V_{Te}$)

The neutralized ion Bernstein wave (NIBW) or the backward branch of the electrostatic ion cyclotron wave was also observed in the ACT-1 neon plasma. This branch was predicted in the 60's when FLR effects were included in ES ICW dispersion relation.[1] Verification of this hot-ion branch was made possible by the creation of a low-collision finite-ion-temperature neon plasma in ACT-1.[27,28] In the experiment, a phased antenna structure immersed in the plasma was used to launch waves with a parallel phase velocity that satisfies the condition for propagation ($V_{Ti} \ll \omega / k_{\parallel} \ll V_{Te}$). In Fig. 11, the measured dispersion relation is shown by the open triangles and dots, and the theoretical solution of the dispersion relation is shown as a solid curve. The part of the dispersion curve that has a negative slope is the backward branch, and the place where the curve turns downward is the mode-conversion point. Experimental data are seen to yield quite good agreement with the theoretical curve.

MODE-TRANSFORMATION AND MODE-CONVERSION PROCESSES

A. Mode-transformation of EPW into IBW ($\omega \approx$ a few Ω_i)

As mentioned above, the EPW $\rightarrow$ IBW mode-transformation process can be used to launch the IBW from an external antenna. Since the IBW is a mode that only exists in a finite ion-temperature plasma, it is important to understand the EPW region that connects the IBW region with the external antenna. The excitation of the EPW by an external antenna has been investigated in detail for the lower hybrid wave heating,[36,37] and this mode-transformation process has been observed and studied in detail in ACT-1 warm hydrogen plasmas.[20,21] By changing the neutral pressure in ACT-1, the ion temperature could be varied over a wide range (1/40 eV to 2 eV), and Fig. 12(a) shows the interferogram output for several neutral pressures, as labeled. For the high-pressure case, the ions are essentially cold. The excited wave is an EPW that stays near the plasma edge, on the low density side of the cold plasma resonance or the lower hybrid resonance [shown for the $T_i \approx 0$ case in Fig. 12(b)]. As the neutral pressure is reduced (T_i increased), a gradual transformation into an IBW is observed. In Fig.12(b), the measured wave number (dots) is plotted as a function of radial

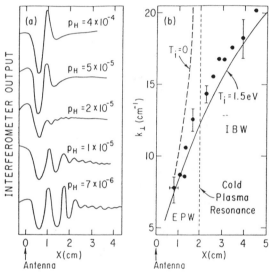

Fig. 12. Verification of EPW-IBW mode transformation in ACT-1.

position in the low-pressure warm-ion plasma. The solid curve is derived from Eq. (1) for $T_i = 1.5$ eV; the dashed curve is for $T_i = 0$. As expected, no sign of discontinuity was observed near the transformation region, the cold plasma resonance, in the experiment as long as T_i was held sufficiently high ($T_i \geq 0.5$ eV).

The transition from EPW ($\omega > \omega_{pi}$) to IBW ($\omega < \omega_{pi}$) can also be seen clearly in Fig. 13 (a), where the wave-packet amplitude (heavier curve) and interferogram output (lighter curve) are shown for various central plasma densities (labeled). The higher central density implies a larger density gradient. In the figure, the radial position of the cold lower-hybrid resonance layer which separates the EPW and IBW regimes is indicated by a dashed curve, and one sees that the wave transition between the two regimes is quite smooth. It is particularly interesting to note here that the IBW was launched effectively even when the ω_{pi} layer almost reached the limiter radius, approaching within a few millimeters of the antenna surface. Figure 13(b) shows the calculated ray position for various $\lambda_{||}$ (as labeled), with corresponding plasma parameters. The position of the wavepacket follows that of a ray for $\lambda_{||} \cong 36$ cm, which is the dominant launched $\lambda_{||}$ in the experiment. The IBW wavepacket is seen to display a dispersive characteristic (spreading of its trajectory due to the spread in $\lambda_{||}$) while the EPW wavepacket shows a resonance-cone characteristic.

A similar launching at the third ion-cyclotron harmonic was also demonstrated in ACT-1. In Fig. 14(a), the measured wave dispersion relation of the excited IBW is shown by the dots. The calculated values are shown by the solid curves for various values of ion temperature. In Fig. 14(b), the measured wavenumber is plotted as a function of the plasma position (dots). As in Fig. 14(b), the theoretical values for $T_i = 2$ eV and $T_i = 0$ eV are shown. Again, a smooth mode-transformation could be seen across the cold LH resonance layer in a good agreement with theory. In this third harmonic launching case, a higher ion temperature ($T_i \approx 2$ eV) was required for efficient IBW launching as expected.

One should note that the external launching of the cold electrostatic ion cyclotron wave (CES ICW) also involves mode transformation of the launched EPW into a CES ICW near the ion plasma frequency. In this case, FLR effects are not important (except when the CES ICW approaches the ω_{ih} resonance). This manner of external CES ICW launching was demonstrated in the L-4 experiment described above.

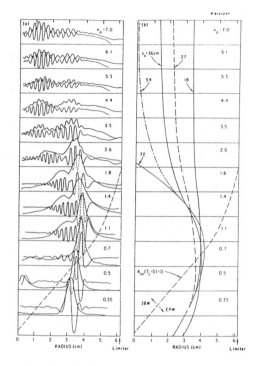

Fig. 13. EPW / IBW wave-packet radial profiles versus plasma density.

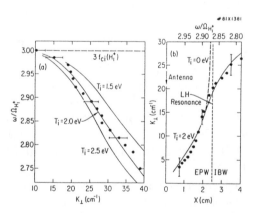

Fig. 14. Third-harmonic launching of IBW in ACT-1.

B. Mode-conversion in the lower hybrid waves ($\omega \gg \Omega_i$)

As the lower hybrid wave or EPW approaches the lower hybrid resonance, the FLR term becomes larger and the mode conversion into hot plasma waves (HPW) takes place. This process, predicted by Stix, can provide a mechanism through which ions are heated near the conversion layer.[3] Previous attempts to detect the mode-conversion process in several experiments were not successful partly due to the relatively cold-ion collisional plasmas. The ACT-1 plasma provided an ideal test bed to explore this long outstanding physics problem. The process nevertheless is complex in that the mode-conversion point for the each $k_{\parallel}$ component is different both radially and toroidally, resulting in a complex interference pattern. In ACT-1, a significant broadening of the resonance cone was observed near the lower-hybrid resonance layer, Fig. 15.[26] The observed interference pattern revealed a periodic structure near the ion cyclotron harmonic frequencies, marking the influence of FLR effects. Modelingthe experiment with a hot-ion electrostatic ray tracing code together with the Fourier reconstruction of the waveforms offered additional confirmation that linear mode conversion into HPW indeed occurred in the experiment.

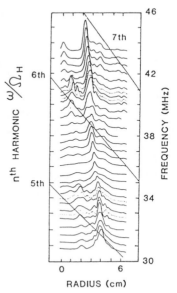

Fig. 15.   Observation of lower-hybrid mode-converted hot-plasma modes in ACT-1.

VI. Application of IBW for plasma heating

In ion Bernstein wave heating (IBWH), a FLR wave, rather than the usual cold plasma wave, is used to transport the rf power to heat the plasma core in a tokamak reactor.[14,25] Careful investigation of the IBW has brought to light unique properties of the wave that are attractive for reactor application. Early wave accessibility studies showed that this FLR mode would be able to penetrate to the hot dense reactor plasma core without significant attenuation. Further investigation revealed additional useful properties. Due to the low phase velocity ($\omega / k_{\perp} \approx V_{Ti} \ll V_{\alpha}$, where V_{α} is the velocity of the fusion alpha particles), wave absorption by the 3.5 MeV fusion α-particles, a potentially serious problem, can be avoided. Also, the FLR wave property that $k_{\perp}\rho_i \approx 1$ makes localized bulk ion heating possible at the ion cyclotron harmonic layers. Such heating of the bulk ion distribution can be desirable for optimizing fusion reactivity. Finally, the EPW-IBW mode transformation permits utilization of a lower-hybrid like waveguide launcher that is compatible for reactor application, as shown in Fig. 16.

The first IBW heating experiment was conducted in the ACT-1 toroidal device.[23] Using probe diagnostics, the physics of IBWH were investigated for all phases of the wave-heating scenario: IBW launching, propagation, absorption and heating.[20,-23] In a hydrogen plasma ($T_e \approx 2.5$ eV, $T_i \approx 1.5$ eV, and $n_0 \leq 10^{11}$ cm^{-3}), detailed profile measurements of the wave absorption and the resulting ion-temperature increase identified the heating layers near the ion cyclotron harmonics of deuterium-like (H_2^+) and tritium-like ions (H_2^+), where the dominant absorption resonances exist at $5\Omega_D$ and $5\Omega_T$.[23] Figures 17 and 18(a) show T_i versus frequency and radial position, confirming that the maximum

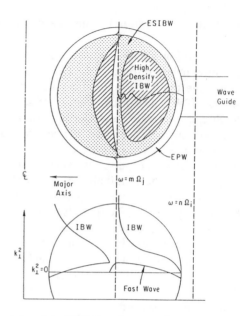

Fig. 16   IBW heating configuration in typical tokamak parameters. (a) IBW waveguide launcher is shown on the left. Plasma cross-section is separated according to the various regimes. Going from low to high density, the electron plasma wave, ion Bernstein wave, and high β ion Bernstein wave regimes, respectively.

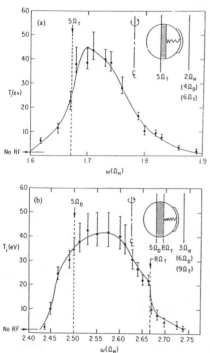

Fig. 17.   Ion temperature versus frequency during IBW heating in ACT-1.

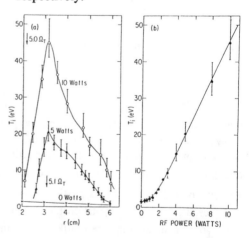

Fig. 18.   IBW heating experiment in ACT-1. (a) Ion temperature profiles for various rf power levels (as labeled). (b) Peak ion temperature versus rf power

ion temperature increase indeed occurred near the heating layer. In this experiment, the antenna phase was chosen to avoid electron Landau damping of the wave. Interestingly, as shown in Fig. 18 (b), the central ion heating efficiency actually increases with rf power, as higher T_i improves wave penetration. In the lower temperature range, collisions caused partial absorption of IBW power near the plasma edge. As the ions are heated, the higher T_i increases the group velocity (since $V_g \propto V_{Ti}$) and reduces the collisional absorption (Im $k_\perp \propto V_{Ti}^{-4}$). In the best case, power balance estimates suggest that the ion heating is nearly 100% efficient for IBWH. Moreover, ion energy analyzer data showed that the heat deposition was into the bulk ions without any significant production of an ion tail. This characteristic of IBWH can be explained by quasilinear-diffusion / Fokker-Planck collision theory[42] in the large FLR limit.[25]

The B_θ-loop was also tested on ACT-1, where the physics of IBWH wave loop-excitation and the antenna loading was investigated in considerable detail.[29,31] After the ACT-1 experiment, IBW heating was investigated in many tokamak devices including JIPPTII-U, TNT, PLT and Alcator-C, JFT-II-M, DIII-D, and PBX-M. The first waveguide IBW experiment is planned on the FT-U tokamak. A review article summarizing the IBW-related activities can be found in Ref. 25.

PARASITIC EXCITATION OF ELECTROSTATIC WAVES

Electrostatic waves can be excited inadvertently in high-power ICRF experiments. The anomalous antenna loading observed during the Faraday-shield-less operation of the Ion Cyclotron Wave Heating (termed Mode X) may be one well known example.[39] Mode X may be the actual external launching of a CES ICW which unfortunately cannot propagate to the plasma core. It was noted earlier that in L-4, various parametric excitations of electrostatic waves during the ICRF were investigated.[9-11] And in ACT-1, two dominant parasitic IBW excitation processes by the near field of the Faraday-shielded ICRF antenna have been observed.[29,30]

Parametric Instabilities - With relatively low threshold electric field, strong parametric instability activity has been observed where the ion Bernstein wave and an ion quasi-mode are excited as the lower sideband and low-frequency mode, respectively.[30] Figure 19 portrays the observed instability amplitude and the calculated growth rate. Agreement between the observed amplitude and the calculated growth rate is quite good, especially considering that the observed amplitude is the nonlinear saturated value of the unstable mode. Due to a quite restrictive selection-rule condition, parametric IBW excitation is not likely to play an important role under normal fast-wave heating conditions. The likelihood of excitation will, however, increase for higher frequency operation ($\omega >> 2\Omega_i$) due to a lower power threshold and to the availability of more decay modes.

Density Gradient Excitation - Ion Bernstein waves can also be excited through a density-gradient-driven process.[29] In Fig. 20, a two-dimensional wave interferogram of the excited IBW is shown. Noting that the excitation efficiency depends on the density gradient in front of the antenna, one may attribute the nonuniform excitation pattern to the nonuniform density gradient along the antenna surface. In the context of the slab model, the E_y field of the fast-wave antenna couples to the IBW E_x field through the spatial derivative term, $\partial_x K_{xy}$ in the density gradient region. A theory based on a local analysis of the Vlasov-Maxwell equations including the effect of the plasma density gradient was developed and found to be in very good agreement with the experimental observation.[29] Such a process can assume increased importance if the antenna loading into the fast wave is not sufficiently high, a condition that may occur if the antenna-plasma gap distance is increased.

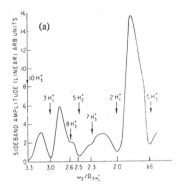

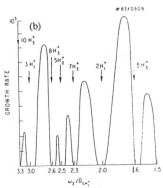

Fig. 19. Parametric excitation of IBW in ACT-1. (a) Measured IBW decay wave amplitude versus ω(decay) / Ω_H. The toroidal field increases to the left. Ion cyclotron harmonic frequencies are indicated. (b) Calculated growth rate for the corresponding experimental parameters.

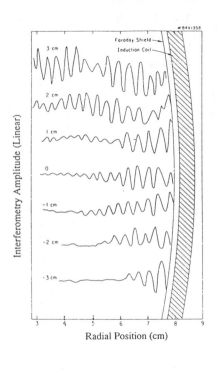

Fig. 20. Parasitic excitation of IBW in ACT-1. The measured ion Bernstein wave interferogram is plotted over the poloidal section of the plasma. In the toroidal direction, the probe is located 34 cm from the Faraday shielded ICRF antenna.

CONCLUDING REMARKS

In conclusion, a variety of electrostatic waves and related processes in the ion cyclotron range of frequencies were investigated in L-4 and ACT-1. Almost every successful experimental measurement showed surprisingly good agreement with the FLR magnetized wave theory developed by Stix,[1] Bernstein,[33] and others. However, this fact certainly did not diminish the satisfaction of actually observing the predicted phenomena for the first time. One is usually unsure until the phenomena are actually observed in an experiment. After confirmation, they can be treated as something obvious. This sequence is, in fact, the important function of experiments. Once waves were well understood in both theory and experiment, they often served as useful tools for plasma diagnostics, e.g., EPW for n_e, ES ICW for T_e, T_i, and n_i, IBW for T_i, and n_i, etc. However, it was not always easy to predict the outcome of an experiment. In dealing with actual experimental situations there are many possible reasons for not observing the expected phenomena. Excessive fluctuations, for example, can mask the phenomena being investigated. Some waves are more susceptible to fluctuations than others. The neutralized IBW was a relatively difficult wave to detect and the reason for the difficulty is not fully understood even today. The search for the mode-converted hot-plasma waves near the lower-hybrid resonance was also quite complicated and required a sophisticated analysis. On the other hand, some experiments turned out to be easier than expected. Experiments to externally excite the CES ICW and IBW are examples. In ACT-1, the IBW was observed almost immediately after the "switch" was turned on. We were especially impressed by the coherence of IBW propagation in ACT-1. And indeed, the effect of fluctuations on IBW was later calculated to be relatively small.[43] Through experimental investigation, we came to appreciate wave absorption processes such as ion cyclotron damping and electron Landau damping. We also became keenly aware of nonlinear processes even when investigating processes that were nominally linear.

Looking to the future, I believe that such basic physics experiments are enormously useful for the main-line fusion program. They are certainly important for the current IBW experiments in large tokamaks. IBW heating, moreover, appears to have many applications in addition to heating fusion plasmas (see Ref. 25 for example). Areas of application for these waves may include astronomical as well as industrial plasmas.

I would like to conclude this paper by adding one more possible application for these waves. For a fusion reactor, impurity control and removal of helium ash are important issues. The ash accumulation, if unchecked, can degrade or even quench the fusion reaction. The helium ash is a noble gas and rather difficult to pump compared to other more reactive gases. There are several ideas for pumping helium ash by selective heating near the plasma edge. As the helium ions are heated in the edge region, they can be trapped by the field ripple (natural or created) which then can cause the trapped helium ions to drift rapidly out of the plasma. A trap of some sort could, presumably, then capture them. This scheme would pump helium both as it comes out from and as it re-enters the plasma. In principle, the scheme can be used for other impurity ions as well. The idea is certainly not new – fast-wave heating has already been suggested for this purpose.[44] However, the selective local absorption of the fast wave by the minority helium ions in the low temperature, low-β edge region may not be efficient. One possible improvement proposed here is to use the externally launched ion Bernstein wave (IBW) or the cold electrostatic ion cyclotron wave (CES ICW). Both electrostatic waves can be used to heat the helium resonance with relatively high efficiency even at the edge. Although not suited for core plasma heating, the CES ICW should work well for such edge heating. External launching of both waves was already demonstrated in the L-4 and ACT-1 experiments. Singly ionized helium ions have a unique ion cyclotron layer and, therefore, may be selectively heated. Moreover, the wave fields of those externally launched electrostatic waves can be localized toroidally. For IBW, one might heat at $3\Omega(^4He)$ using $2\Omega_D$ launching or at $5\Omega(^4He)$ using $4\Omega_T$ launching. For the CES ICW, an obvious choice is $\Omega(^4He)$ heating using Ω_T launching. It should be noted here that the FLR modification of the CES ICW in the region between the tritium-helium ion-ion hybrid frequency and the helium ion-cyclotron frequency would be expected to improve the power absorption. Even though the required ion energy for helium removal is not known precisely (but probably in the range of few hundred to few keV), the longer wavelength of the CES ICW could well result in heating the helium ions to higher energy compared to IBW. This manner of ash removal can be investigated in the present-day tokamaks using existing IBW antennas (such as in PBX-M), since the IBW-type antenna is well suited to launch both the IBW and the CES ICW.

Acknowledgments: The author is deeply indebted to Dr. Stix' teaching of waves through his class, his books and manuscripts, and frequent private discussions. He also thanks him for giving much needed guidance, support, and encouragement over years for the work described here. The author would like as well to express his appreciation to his former Ph. D. advisors, Drs. M. Porkolab and R. P. H. Chang. He also thanks Dr. K. L. Wong for his contributions to the work presented here through the ACT-1 device construction and subsequent experiments. This work also describes the doctoral thesis work performed by the former ACT-1 students, Drs. G. A.Wurden, F. N. Skiff, and J. A. Goree. This work was supported by U.S. Department of Energy contract No. DE-AC02-76-CHO-3073.

REFERENCES

1. T. H. Stix, Theory of Plasma Waves (McGraw-Hill, New York 1962). See also Waves in Plasmas (AIP, New York 1992).
2   D. G. Swanson and Y.C. Nygan, Phys. Rev. Lett. 35, 517 (1975).
3. T. H. Stix, Phys. Rev. Lett. 15, 878 (1965).
4. A. Y. Wong, N. D'Angelo, and R. W. Motley, Phys. Rev. Lett. 9, 415 (1962).
5. E. R. Ault and H. Ikezi, Phys. Fluids 13, 2874 (1970).
6. J. P. M. Schmitt, Phys. Rev. Lett. 31, 982 (1973).
7. S. J. Buchsbaum, Phys. Fluids, 3,418 (1960)
8. J. L. Sperling and F. W. Perkins, Phys. Fluids 17, 1857 (1974).
9. M. Ono, Ph.D. Thesis, Princeton University (1978).
10. M. Ono, M. Porkolab, and R.P.H. Chang, Phys. Rev. Lett.. 38, 962 (1977).
11. M. Ono, M. Porkolab, and R.P.H. Chang, Phys. Fluids 20, 1656 (1980), and 20, 1675(1980).

12. M. Ono, Phys. Rev. Lett. 42, 1267 (1979).
13. N. J. Fisch, Phys. Rev. Lett. 41, 873 (1978).
14. See National Technical Information Service Document No. PPPL-1593 ("Plasma heating by externally launched ion Bernstein waves," by M. Ono, 1979). Copies may be ordered from NTIS, Springfield, Virginia 22161; Phys. Fluids 28, 2645 (1985).
15. K. L. Wong, M. Ono, and G.A. Wurden, Rev. Sci. Instrum. 53(4), 409 (1982).
16. K. L. Wong, R. Horton, and M. Ono, Phys. Rev. Lett. 45, 117 (1980).
17. J. Goree, M. Ono, P. Colestock, R. Horton, D. McNeill, and H. Park, Phys. Rev. Lett. 55, 1669 (11985).
18. M. Ono, G. J. Greene, D. S. Darrow, C. Forest, H. Park, and T. H. Stix, Phys. Rev. Lett. 59, 2165 (1987).
19. C. B. Forest, Y. S. Hwang, M. Ono, and D. S. Darrow, Phys. Rev. Lett. 68, 3559 (1982).
20 M. Ono, K.L. Wong, Phys. Rev. Lett. 45, 1105 (1980).
21 M. Ono, K. L. Wong, and G. A. Wurden, Phys. Fluids 26, 298 (1983).
22 G. A. Wurden, M Ono, and K. L. Wong, Phys. Rev. A 26, 2297 (1982).
23 M. Ono, G. A. Wurden, and K. L. Wong, Phys. Rev. Lett. 52, 37 (1984).
24 G. A. Wurden, Ph.D. Thesis, Princeton University (1982).
25 M. Ono, Phys. Fluids B, Feb. (1993). A review article on IBW heating.
26. G. A. Wurden, K. L. Wong, F. Skiff, and M. Ono, Phys. Rev. Lett. 50, 1779 (1983).
27 J. A. Goree, Ph.D. Thesis, Princeton University (1985).
28 J. A. Goree, M. Ono, and K. L. Wong, Phys. Fluids 28, 2845 (1985).
29. F. N. Skiff, Princeton University, Ph.D. Thesis (1985).
30 F. N. Skiff, M. Ono, and K. L. Wong, Phys Fluids 27, 1051 (1984).
31. F. N. Skiff, M. Ono, P. Colestock, and K. L. Wong, Phys. Fluids 28, 2453
32. B. D. Fried and S. D. Conte, The Plasma Dispersion Function (Academic, New York, 1961).
33. I. B. Bernstein, Phys. Rev. 109, 10 (1958).
34. P. K. Fisher and R. W. Gould, Phys. Fluids 14, 857 (1971).
35. H. H. Kuel, Phys. Fluids 17, 1636 (1974).
36. P. M. Bellan and M. Porkolab, Phys. Fluids 17, 1592 (1974), and 19, 995 (1976).
37. S. Bernabei, M. A. Heald, W. H. Hooke, R. W. Motley. F. J. Paoloni,. M. Brambilla, and W. D. Getty, Nucl. Fusion 17, 929 (1977).
38. D. G. Swanson, Phys. Fluids 10, 1531 (1975).
39. M. A. Rothman, R. M. Sinclair, and S. Yoshikawa, J. Nuclear Energy, Part C, 8, 241 (1966).
40. D. G. Swanson, , Phys. Fluids 28, 2645 (1985).
41. R. Limpacher and K. R. Mackenzie, Rev. Sci. Instrument 44, 726 (1973).
42. T. H. Stix, Nucl. Fusion 15, 753 (1976).
43. M. Ono, Phys. Fluids 25, 990 (1982).
44. J. C. Hosea, Private Communication.

THE MAGNETOHYDRODYNAMIC RANKINE-HUGONIOT RELATIONS

Charles F. Kennel
University of California, Los Angeles, CA 90024-1547

1.   INTRODUCTORY REMARKS

A compressional simple wave will evolve until dissipation balances steepening to form a shock. The localized dissipation layer cannot be treated by ideal magnetohydrodynamics, which conserves entropy along flow streamlines. However, the stationary states asymptotically far upstream and downstream of the shock are free of dissipation because they are spatially uniform. Consequently, ideal MHD does describe the changes in flow parameters between the two stationary states, provided that entropy conservation is relaxed.

The MHD Rankine-Hugoniot conditions relate flow states with the same electric field and mass, momentum, and energy fluxes. They are obtained by substituting energy conservation for entropy conservation and integrating the resulting set of MHD conservation laws across the shock[1]. The Rankine-Hugoniot relations do not depend upon the nature of the dissipation processes in the thin shock layer, provided that their net effect is to restore local thermodynamic equilibrium.

Because ideal MHD is scale-free, it permits us to imagine that the upstream and downstream states will be separated by a thin discontinuity. Rigorously speaking, this conclusion requires the existence of a shock structure solution that becomes thin in the limit of small dissipation. However, having noted this caveat, we will continue the common usage and call the transitions predicted by the Rankine-Hugoniot relations "discontinuities".

Section 2 outlines general properties of the magnetohydrodynamic Rankine-Hugoniot relations. The complete set of jump relations is obtained in Section 2.1 by integrating the conservation laws from upstream to downstream. In Section 2.2, we isolate the one discontinuity that does not propagate with respect to the fluid. The entropy discontinuity is simply a non-propagating boundary separating two fluids which have different densities, temperatures, and entropy densities. The remaining MHD discontinuities propagate with respect to the fluid upstream.

In Section 2.3, we eliminate the entropy discontinuity from the Rankine-Hugoniot relations, and arrive at a set of equations describing propagating discontinuities. Specification of the discontinuity speed reduces the number of independent variables in ideal MHD from seven to six. However, three of these variables may be removed by a Galilean transformation to the normal incidence frame of reference and an azimuthal rotation of the coordinate system. Consequently, the jump relations reduce to three equations in three unknowns, which we choose to express in the forms, $F(r, \mathbf{b_T}) = 0$ and $\mathbf{Z}(r, \mathbf{b_T}) = 0$, where $r = V_x/V_{x1}$ is the normalized velocity contrast across the discontinuity, and $\mathbf{b_T}$ is the magnetic field two-vector transverse to the direction of propagation. This form of the conservation laws expresses the invariance of the one-dimensional ideal MHD equations to rotations about the direction of propagation.

The relation $F(r, \mathbf{b_T}) = 0$ determines a surface, or surfaces, in the three-dimensional $(r, \mathbf{b_T})$ space on which the fluxes of mass, momentum, and energy are the same. The two equations $\mathbf{Z}(r, \mathbf{b_T}) = 0$ define surfaces on which the shock-frame electric field is the same, and express the conservation of the magnetic flux transported across the discontinuity by the moving fluid. The Rankine-Hugoniot relations for propagating discontinuities consist of the simultaneous solution of $F = 0$ and $\mathbf{Z} = 0$.

We elucidate the relationship between the magnetohydrodynamic and hydrodynamic Rankine-Hugoniot relations in Section 2.4. Of course, MHD reduces to hydrodynamics if we take the magnetic field to be zero. But the connection is even deeper. It turns out that the solution to $F(r,\mathbf{b_T}) = 0$ has two branches which are the states upstream and downstream of a hydrodynamic shock at each value of $\mathbf{b_T}$. This property of $F(r,\mathbf{b_T})$ will prove central only to our discussion of the MHD Rankine-Hugoniot relations.

We introduce the important distinction between parallel and oblique propagation in Section 3. The existence of a component of transverse magnetic field upstream of obliquely propagating discontinuities makes it possible to define a so-called coplanarity plane, the one containing the upstream magnetic field and the direction of propagation. The relations $\mathbf{Z}(r,\mathbf{b_T}) = 0$ then have the following implications. The downstream magnetic field either remains in or rotates out of the coplanarity plane. If it rotates out, the speed of the discontinuity must equal the upstream intermediate speed. Conversely, if the speed of the discontinuity does not equal the intermediate speed upstream, it is coplanar.

We conclude Section 3 by showing that the noncoplanar discontinuity which propagates at the intermediate speed rotates the directions of the transverse magnetic field and fluid velocity without changing any other property . For this reason, it has been given the name, rotational discontinuity. It leaves the density, pressure, temperature, and the magnitude of the magnetic field unchanged. Its strength, the angle through which the transverse field rotates, is unrelated to its speed. Because it does not change the entropy, it should not really be considered a shock. The rotational discontinuity is really just a discontinuous version of an intermediate simple wave.

It is possible to factor the rotational discontinuity out of the Rankine-Hugoniot relations to obtain jump relations for oblique <u>coplanar</u> discontinuities whose speeds differ from the intermediate speed. This reduces the 3 equations, $F(r,\mathbf{b_T}) = 0$ and $\mathbf{Z}(r,\mathbf{b_T}) = 0$, to two, $F(r,b) = 0$, and $Z(r,b) = 0$, where b is now the component of magnetic field in the coplanarity plane, and Z is the y-component of $\mathbf{Z}$. By eliminating the magnetic field in favor of the velocity contrast,r, we combine these two equations in Section 4 into a single quartic equation in r whose solution selects at most four real Rankine-Hugoniot states. It is illuminating to reexpress this quartic by replacing the velocity contrast by an equivalent independent variable, X, whose sign measures whether the discontinuity speed is above or below the intermediate speed.

The MHD Rankine-Hugoniot relations are invariant with respect to a Galilean transformation to a frame moving with an arbitrary velocity transverse to the direction of propagation. Two particular frames of reference are commonly used. In the standard normal incidence frame, we assume that the transverse velocity is zero upstream. DeHoffman and Teller showed that the electric field can be removed altogether by a Galilean transformation. The fact that the fluid velocity in the so-called DeHoffman-Teller frame is parallel to the magnetic field both upstream and downstream simplifies the form of the Rankine-Hugoniot relations, as we show in concluding Section 4.

However, the DeHoffman-Teller transformation cannot be used for perpendicular shocks, and we prefer to work in the standard normal incidence frame.

We classify the possible MHD shocks in Section 5. We first number the four possible coplanar Rankine-Hugoniot states, 1, 2, 3, and 4, in order of increasing density and decreasing velocity contrast. Then, by examining the properties of the curves $F(r,b) = 0$, and $Z(r,b) = 0$ near their points of intersection, we show that the Rankine-Hugoniot states may be classified according to the relation between the normal component of the flow speed and the MHD characteristic speeds in each state. The flow speed exceeds the fast speed in state 1, and is between the fast and intermediate speeds in state 2. It is between the intermediate and slow speeds in state 3, and is below the slow speed in state 4.

The MHD shocks are transitions between pairs of Rankine-Hugoniot states whose order is determined by the requirement that the entropy density downstream exceed that upstream. The allowed coplanar shocks are six in number and are all compressional. They are of types 1-2, 1-3, 1-4, 2-3, 2-4, and 3-4, where our convention is to label the upstream state first and the downstream state second. These six shocks may be divided into two types according to the direction of the downstream transverse magnetic field. The transverse field has the same direction upstream and downstream of the 1-2 and 3-4 shocks. These we call fast and slow shocks respectively, because they reduce to the fast and slow MHD waves in the small amplitude limit. The magnetic field rotates by 180 degrees across the remaining four discontinuities, which are called intermediate shocks, because they take the normal component of the flow speed from above to below the intermediate speed. Intermediate shocks have no small amplitude analogs, since a 180 degree field rotation cannot be considered small.

In sum, one-dimensional MHD has seven independent variables, and, counting the entropy discontinuity, the MHD Rankine-Hugoniot relations allow seven possible shocks across which the entropy changes.

The above definitions of fast, slow and intermediate shocks are unambiguous when the coplanarity plane can be defined on both sides of the shock. In section 5.3, we discuss switch-on and switch-off shocks, which emerge when the coplanarity plane is undefinable on one side or other of the shock. In these cases we must return to the full three dimensional Rankine-Hugoniot relations, $F(r,\mathbf{b_T}) = 0$ and $\mathbf{Z}(r,\mathbf{b_T}) = 0$. The switch-on fast shock propagates parallel to the upstream magnetic field with the downstream intermediate speed. It "switches on" a transverse component of magnetic field downstream, whose azimuthal orientation is arbitrary. The switch-off slow shock propagates parallel to the downstream magnetic field at the upstream intermediate speed, and "switches off " the upstream transverse field.

We elaborate upon the special case of parallel propagation in Section 5.4. The switch-on shock exists only in the finite range of speeds for which the Rankine-Hugoniot relations have four real solutions. In this case, the possible shocks are the switch-on and switch-off shocks, and a 1-4 intermediate shock which satisfies the jump condition for a hydrodynamic shock. Outside this range, the only remaining solution is the hydrodynamic shock, which is classified as a 1-2 fast shock when the Alfven speed exceeds the sound speed upstream, and a 3-4 slow shock, when the sound speed exceeds the Alfven speed upstream.

The relationship between switch-on shocks, the rotational discontinuity, and the intermediate shocks is clarified in Section 5.5. The intermediate shocks exist only for the limited range of flow parameters for which there exist four real Rankine-Hugoniot solutions. By calculating where two of the four solutions turn complex, we show that intermediate shocks have speeds between the intermediate speed and an upper limit which reduces to the switch-on upper limit speed for parallel propagation. The parameter range in which intermediate shocks occur conveys a clear message: intermediate shocks break the azimuthal symmetry of the one-dimensional equations expressed by the switch-on shock and the rotational discontinuity. However, when the upstream transverse field which breaks the symmetry is too large, we obtain only a fast shock or a slow shock.

Not all the possible combinations of Rankine-Hugoniot states will necessarily be realized as shocks. To decide which ones will actually occur, one has to invoke additional conditions derived from physical reasoning beyond that contained in the formulation of the conservation laws. We already used an entropy condition to select six thermodynamically admissible shocks from the twelve possible combinations of different Rankine-Hugoniot states. Furthermore, we are only interested in those shocks that evolve from smooth initial conditions, and can adjust to variable external boundary conditions. The conditions imposed by such issues of dynamical causality, the so-called MHD evolutionary conditions, are discussed in Section 6.

There never has been any doubt about the evolutionarity of the fast and slow shocks. However, evolutionarity arguments have suppressed all consideration of intermediate shocks for the past quarter century. We summarize the arguments against intermediate shocks, point out that stable intermediate shocks do evolve in numerical integrations of the MHD Navier-Stokes equations, and indicate several possible weaknesses in the formulation of the MHD evolutionary conditions.

In Appendix A, we survey the dependence of the properties of fast, intermediate and slow shocks upon their speeds, and the direction and strength of the upstream magnetic field. Although none of this information is truly fundamental, all who work experimentally on shock waves in plasmas are acutely aware of their parameter sensitivity.

2.   PROPERTIES OF THE MHD RANKINE-HUGONIOT RELATIONS

2.1.   Complete set of jump relations

Let us consider a plane discontinuity of finite amplitude propagating in the positive x-direction into a uniform stationary upstream state. We transform to the fram of reference moving with the discontinuity, and integrate the equations of ideal magnetohydrodynamics from upstream to downstream. In this frame, the equation of mass conservation integrates to

$$[\rho V_x]^* = \rho V_x - \rho_1 V_{x1} , \qquad (1a)$$

where subscript 1 denotes the upstream state, and the starred bracket signifies the jump across the discontinuity. Unsubscripted variables refer to the downstream state. Using the facts that ρV_x and B_x are conserved, we may integrate the equations of momentum conservation to obtain

$$\left[\rho V_x^2 + P + \frac{B_T^2}{4\pi}\right]^* = 0 , \left[\rho V_x \mathbf{V_T} - \frac{B_x \mathbf{B_T}}{4\pi}\right]^* = 0 . \qquad (1b,c)$$

Since entropy is not necessarily conserved across finite amplitude discontinuities, we replace entropy conservation with energy conservation. When it is integrated across the discontinuity, the energy conservation relation takes the form

$$\left[\rho V_x \left(\frac{V_x^2 + V_T^2}{2}\right) + \frac{\gamma}{\gamma - 1}\frac{V_x P}{}\right]^* + \frac{c}{4\pi}[(\mathbf{E} \times \mathbf{B})]_x)^* = 0 , \qquad (1d)$$

where E is the electric field.

For stationary discontinuities, Faraday's law requires that $\nabla \times \mathbf{E} = 0$, which in one dimension reduces to

$$[\mathbf{E_T}]^* = 0 , \qquad (1e)$$

where $\mathbf{E_T} = (0, E_y, E_z)$ is the transverse electric field. Since no charge can accumulate in a steady discontinuity, integration of Poisson's law implies $[E_x]^* = 0$.

Our final requirement is that the system satisfy the MHD Ohm's law, $\mathbf{E} + \mathbf{V} \times \mathbf{B}/c = 0$, both upstream and downstream, so that (1e) becomes

$$[V_x \mathbf{B_T} - \mathbf{V_T} B_x]^* = 0 . \qquad (1f)$$

2.2.   Entropy and tangential discontinuities

Let us first consider non-propagating solutions with $V_{x1} = 0$. Equation (1a) then implies that $V_x = 0$ (for nonzero ρ) so that the fluid moves parallel to the surface of discontinuity on either side of it. Then, since $\rho V_x = 0$, (1b), (1c), and (1g) reduce to

$$\left[P + \frac{B_T^2}{8\pi}\right]^* = 0 , B_x[\mathbf{B_T}]^* = 0 , B_x[\mathbf{V_T}]^* = 0 , \qquad (2a,b,c)$$

and since $[E]* = 0$, (1d) becomes

$$\frac{c}{4\pi} \mathbf{E_1} \times [\mathbf{B_T}]* = 0 . \qquad (2d)$$

Equation (2d) is then trivially satisfied.

Equations (2a-d) describe the entropy discontinuity, a boundary separating two fluids in dynamical equilibrium which have different densities, temperatures, and entropies. This is sometimes called the contact discontinuity. When B_x is non-zero, equations (2b) and (2c) imply that the tangential velocity and magnetic field are continuous, and, together with (2a), that the pressure is continuous across the discontinuity.

The special case where B_x and V_x are both zero is less constrained. Here, equations (1a), (1c), (1d) and (1g) are satisfied identically. The flow velocity and magnetic field are both parallel to the surface of discontinuity. They can have any jump in direction and magnitude allowed by (1f), and the density can change by an arbitrary amount, subject to the condition of magnetohydrostatic force balance, (2a). Because of its resemblance to the tangential discontinuity in ordinary hydrodynamics, the $B_x = 0$ entropy discontinuity is also called the MHD tangential discontinuity.

2.3.   Rankine-Hugoniot relations for propagating discontinuities

Henceforth, we will assume that $V_{x1} \neq 0$. Let us now define a convenient set of normalized variables

$$r \equiv \frac{V_x}{V_{x1}} , \qquad \mathbf{b} \equiv \frac{\mathbf{B}}{B_1} = (b_x, \mathbf{b_T}) , \; \mathbf{U_T} \equiv \frac{\mathbf{V_T}}{V_{x1}} , \qquad (3a,b,c)$$

and upstream Alfvén, intermediate, and sonic Mach numbers

$$M_{A1}^2 \equiv \frac{4\pi\rho_1 V_{x1}^2}{B_1^2} , M_{I1}^2 \equiv \frac{M_{A1}^2}{b_x^2} = \frac{4\pi\rho_1 V_{x1}^2}{B_x^2} , \quad M_{s1}^2 \equiv \frac{V_{x1}^2}{C_{s1}^2} . \qquad (4a,b,c)$$

Then, using the facts that ρV_x and b_x are constants, and the definitions above, the equations for momentum conservation become

$$r - 1 + \frac{P - P_1}{\rho_1 V_{x1}^2} + \frac{b_T^2 - b_{T1}^2}{2M_{A1}^2} = 0 , \qquad (5a)$$

and

$$\mathbf{U_T} - \mathbf{U_{T1}} = \frac{b_x (\mathbf{b_T} - \mathbf{b_{T1}})}{M_{A1}^2} . \tag{5b}$$

We evaluate the energy conservation equation assuming the electric field satisfies the MHD Ohm's law upstream, whereupon

$$\frac{r^2 - 1}{2} + \frac{U_T^2 - U_{T1}^2}{2} + \frac{\gamma r}{\gamma - 1}\left(\frac{P - P_1}{\rho_1 V_{x1}^2}\right) + \frac{\gamma}{\gamma - 1}\frac{(r - 1)}{M_{s1}^2}$$

$$+ \frac{\mathbf{b_{T1}} \cdot (\mathbf{b_T} - \mathbf{b_{T1}}) - b_x \mathbf{U_{T1}} \cdot (\mathbf{b_T} - \mathbf{b_{T1}})}{M_{A21}^2} = 0. \tag{5c}$$

By substituting (5b) into (5c), we obtain

$$\frac{r^2 - 1}{2} + \frac{\gamma r}{\gamma - 1}\left(\frac{P - P_1}{\rho_1 V_{x1}^2}\right) + \frac{\gamma(r - 1)}{(\gamma - 1) M_{s1}^2} + \frac{(b_T^2 - b_{T1}^2) - Y(\mathbf{b_T} - \mathbf{b_{T1}})^2}{2M_{A1}^2} \tag{6a}$$

where

$$Y = 1 - \frac{1}{M_{I1}^2} . \tag{6b}$$

Finally, we may substitute (5a) into (6a) to obtain $F(r, \mathbf{b_T}) = 0$, where

$$F = Ar^2 + Br + C = 0 , \tag{7a}$$

$$A = -\frac{1}{2}\left(\frac{\gamma + 1}{\gamma - 1}\right) , \tag{7b}$$

$$B = \frac{1}{(\gamma - 1) M_{S1}^2} + \frac{\gamma}{\gamma - 1}\left(1 - \frac{b_T^2 - b_{T1}^2}{2M_{A1}^2}\right) , \tag{7c}$$

$$C = -\frac{1}{2} - \frac{1}{(\gamma - 1) \; M_{S1}^2} + \frac{(b_T^2 - b_{T1}^2) - Y(b_T^2 - 2\mathbf{b}_T \cdot \mathbf{b}_{T1} + b_{T1}^2)}{2M_{A1}^2} \qquad (7d)$$

The relation $F(r, \mathbf{b}_T) = 0$ defines a surface or surfaces in (r, b_y, b_z) space on which the ideal MHD mass, momentum, and energy fluxes are all simultaneously the same, and equal to their values at an arbitrary "upstream" point where the ideal MHD Ohm's law is obeyed. The upstream and downstream solution both lie on the $F = 0$ surface.

The final Rankine-Hugoniot condition, that the MHD Ohm's law (1g) be obeyed downstream, is

$$r\mathbf{b}_T - \mathbf{b}_{T1} = b_x(\mathbf{U}_T - \mathbf{U}_{T1}) \qquad (8a)$$

in the resent notation. Inserting (5b) into (8a), we obtain

$$\mathbf{Z}(r, \mathbf{b}_T) \equiv \mathbf{b}_T X - \mathbf{b}_{T1} Y = 0, \quad X \equiv r - \frac{1}{M_{I1}^2} . \qquad (8b,c)$$

The equations $\mathbf{Z}(r, \mathbf{b}_T) = 0$ define surfaces in (r, b_y, b_z) space on which the ideal MHD Ohm's law is obeyed.

After specification of the shock speed, the Rankine-Hugoniot conditions relate six flow variables in the upstream and downstream state. However, three of these variables may be removed by transformation -- a Galilean transformation to the normal incidence frome in which $\mathbf{V}_{T1} = 0$ plus a rotation that places $\mathbf{b}_{T1}$ along one coordinate axis. Thus, the Rankine-Hugoniot relations for propagating discontinuities consist of the simultaneous solutions of the equations $F(r, \mathbf{b}_T)$ and 0 and $\mathbf{Z}(r, \mathbf{b}_T) = 0$, a set of three equations in the three unknowns. Since V_{x1} enters (7) and (8) only as V_{x1}^2, r will be independent of the sign of V_{x1}, which implies that discontinuities of a given type propagating in the positive and negative x-directions are otherwise identical. The fact that $\mathbf{V}_{T1}$ disappeared entirely from (7) and (8) though it was retained throughout their derivations reflects the Galilean invariance of one-dimensional MHD to translations in any transverse direction.

2.4.   Relationship to hydrodynamic Rankine-Hugoniot relations

When we rewrite $F(r, \mathbf{b}_T)$ in a form that isolates its magnetic field dependences,

$$- 2F(r, \mathbf{b}_T) = (r - 1) H(r) + \frac{\hat{J} (r, \mathbf{b}_T)}{M_{A1}^2} , \qquad (9a)$$

where

$$\hat{J}(r, \mathbf{b_T}) = (b_T^2 - b_{T1}^2)\left(\frac{\gamma r}{\gamma - 1} - 1\right) + Y(b_T^2 - 2\mathbf{b_T} \cdot \mathbf{b_{T1}} + b_{T1}^2), \quad (9b)$$

the field free part leads to the Rankine-Hugoniot jump condition for a hydrodynamic shock, $(r - 1) H(r) = 0$, where

$$H(r) = \frac{\gamma + 1}{\gamma - 1}\left\{r - \frac{\gamma - 1}{\gamma + 1} - \frac{2}{(\gamma + 1) M_{s1}^2}\right\}. \quad (9c)$$

Thus, the condition $F = 0$ reduces to the hydrodynamic Rankine-Hugoniot relation in the unmagnetized limit for which $M_{A1}^2 \to \infty$. But equation (9) has a more profound meaning, which requires two preparatory remarks to explain. First of all, the "upstream" point labeled by subscript 1 is entirely arbitrary and may be chosen to be any point on the surface $F(r, \mathbf{b_T}) = 0$. Second, this surface has two branches, given by

$$r(\mathbf{b_T}) = -\frac{B \pm \sqrt{B^2 - 4A C}}{2A}. \quad (10)$$

Let these two solutions be called $r(\mathbf{b_T})$ and $r'(\mathbf{b_T})$, and let $r \geq r'$. Now, choose any point on the upper surface $r(\mathbf{b_T})$, and consider it to be the "upstream" point. Since $\mathbf{b_T} = \mathbf{b_{T1}}$, $F(r, \mathbf{b_T}) = 0$ reduces to the hydrodynamic relation $(r - 1) H(r) = 0$ on $r(\mathbf{b_T})$. For the same $\mathbf{b_T}$, the point on the lower surface r' is also relate to r by the hydrodynamic Rankine-Hugoniot condition. Thus, for the same $\mathbf{b_T}$, the upper and lower branches of $F(r, \mathbf{b_T}) = 0$ can be connected by a hydrodynamic shock. It follows that $M_s^2 \geq 1$ on $r(\mathbf{b_T})$ and $M_s^2 \leq 1$ on $r'(\mathbf{b_T})$. Furthermore, if $r' = r$, M_s^2 must equal unity, for then

$$\sqrt{B^2 - 4AC} = 0, \quad 2A r + B = 0. \quad (11a,b)$$

Using Bernoulli's law, (5a) it is straightforward to compute the local sonic Mach number,

$$M_s^2 = \frac{\rho V_x^2}{\gamma P}, \quad (11c)$$

on $F(r, \mathbf{b_T}) = 0$ and show that

$$2Ar + B = - \frac{r(1 - 1/M_s^2)}{\gamma - 1} , \tag{11d}$$

so that if $r' = r$, $M_s^2 = 1$. Finally, since the entropy increases across a compressional hydrodynamic shock, the entropy on the lower branch of $F(r, \mathbf{b_T})$ is larger than it is on the upper branch, at the same $\mathbf{b_T}$.

3.   COPLANARITY AND THE ROTATIONAL DISCONTINUITY

We begin by noting that the intermediate Mach number depends only on r,

$$M_1^2 = \frac{4\pi\rho V_x^2}{B_x^2} = r\, M_{11}^2 \tag{12}$$

everywhere in $(r, \mathbf{b_T})$ space. Use of (12) then reduces to (8b) to

$$(M_I^2 - 1)\, \mathbf{b_T} = (M_{11}^2 - 1)\, \mathbf{b_{T1}} , \tag{13a}$$

which is useful to reexpress in component form in the standard incidence frame, in which $b_{z1} = 0$,

$$(M_I^2 - 1)\, b_y = (M_{11}^2 - 1)\, b_{y1} , \quad (M_I^2 - 1)\, b_z = 0 . \tag{13b,c}$$

Equation (13c) restricts the circumstances in which bz can be nonzero downstream. Either bz is zero downstream, or the normal flow speed equals the intermediate speed downstream. Any discontinuity with nonzero b_z satisfies $V_x^2 = C_I^2$ downstream; (13b) then states that either b_{y1} is zero or $V_{x1}^2 = C_{11}^2$ upstream. In other words, either the discontinuity propagates parallel or antiparallel to the upstream magnetic field, or the flow speed equals the intermediate speed upstream as well as downstream. We will discuss parallel propagating discontinuities separately. The oblique discontinuity thatmoves with the upstream intermediate speed will be called the rotational discontinuyity for reasons that will become apparent shortly. Since its upstream and downstream speeds both equal the intermediate speed, (12) implies that r = 1, and mass continuity further implies that $\rho = \rho_1$ downstream. Substituting these conditions into our original equations, we find

$$[\mathbf{V_T}]^* = \frac{[\mathbf{B_T}]}{\sqrt{4\pi\rho_1}} \text{sign } V_{x1} , \tag{14}$$

so that its jump in tangential velocity is an Alfvén velocity based upon the jump in tangential magnetic fields and the upstream density.

The remaining properties of the rotational discontinuity may be found as follows. Substituting $\rho = \rho_1$, $V_x = V_{x1}$ into (1b) yields

$$\left[P + \frac{B_T^2}{8\pi} \right]^* = 0, \tag{15}$$

so that the total pressure is unchanged. We may also rewrite (1d) in the following form, by adding and subtracting $P - P_1$,

$$\frac{V_x[P]^*}{\gamma - 1} + V_x \left[P + \frac{B_T^2}{8\pi} \right]^{*2} + \frac{\rho V_x}{2} \left[V_T - \frac{B_T \, \text{sign} \, V_{x1}}{\sqrt{4\pi\rho}} \right]^* = 0. \tag{16}$$

The second and third terms of (16) are zero by virtue of (14) and (15), so that, using (15) again, we find

$$[P]^* = \left[\frac{B_T^2}{8\pi} \right]^* = 0, \tag{17}$$

so that the magnitude of the magnetic field and the pressure do not change.

The rotational discontinuity rotates the directions of the transverse magnetic field and fluid velocity without changing any other fluid property. Its strength, the angle through which B_T rotates, is arbitrary and unrelated to its propagation speed, which is always the intermediate speed. Since it does not change the entropy, it should not be considered a shock. The intermediate discontinuity is, however, directly related to the corresponding isentropic simple wave. In Chapter 3, we showed that the intermediate simple wave does not steepen and that it may have an arbitrary profile, including a discontinuous one. It turns out that we derived the properties of the discontinuous simple wave using jump relations.

4.   RANKINE-HUGONIOT RELATIONS FOR OBLIQUE COPLANAR DISCONTINUITIES

4.1.   Standard normal incidence frame

If the downstream speed does not equal the intermediate speed, so that $X = r - 1/M_{11}^2 \neq 0$, the Ohm's law condition $Z(r, b_T) = 0$ implies that

$$\mathbf{b_T} = \frac{Y}{X}\, \mathbf{b_{T1}} , \tag{18}$$

which indicates that the downstream tangential magnetic field is parallel or anti-parallel to the upstream tangential magnetic field. for such coplanar discontinuities, it is useful to transform to the standard normal incidence frame ($\mathbf{b_{T1}} = \hat{e}_y b_{y1} = \hat{e}_y \sin\theta_1$, $\mathbf{U_{T1}} = 0$), in which case the Rankine-Hugoniot relations reduce to a pair of algebraic equations, $F(r, b) = 0$, $Z(r, b) = 0$, where we may now define $b = B_y/B_1$, and

$$F = Ar^2 + Br + C = 0 , \tag{19a}$$

$$A = -\frac{1}{2}\left(\frac{\gamma + 1}{\gamma - 1}\right), \tag{19b}$$

$$B = \frac{1}{(\gamma - 1)\, M_{S1}^2} + \frac{\gamma}{\gamma - 1}\left(1 - \frac{b^2 - \sin^2\theta_1}{2M_{A1}^2}\right), \tag{19c}$$

$$C = -\frac{1}{2} - \frac{1}{(\gamma - 1)\, M_{S1}^2} + \frac{(b^2 - \sin^2\theta_1) - Y(b - \sin\theta_1)^2}{2M_{A1}^2}, \tag{19d}$$

$$Z = bX - Y \sin\theta_1 = 0 . \tag{19e}$$

The angle θ_1 between the propagation direction and the upstream magnetic field is called the shock normal angle.

The coplanar Rankine-Hugoniot relations consist of the simultaneous solution of $Z(r, b) = 0$ and $F(r, b) = 0$, which now may be accomplished graphically by plotting both curves in the (r, b) plane and determining their points of intersection. Each curve has two branches, the two solutions of the quadratic $F(r, b) = 0$, and the two hyperbolae defined by $Z(r, b) = 0$. Thus, there can be at most four points of intersction for b and r real.

By specializing (9a) to the coplanar case, and substituting into it the solution, below, to $Z(r, b) = 0$,

$$b = \frac{Y \sin\theta_1}{X}, \quad X = r - \frac{1}{M_{I1}^2}, \tag{21a,b}$$

we obtain an algebraic form of the Rankine-Hugoniot relation,

$$-F\left(X, \frac{Y \sin\theta_1}{X}\right) = \frac{(X - Y)}{2X^2} R(X) = \frac{(r - 1) R(X)}{2X^2} = 0 , \tag{22}$$

where the MHD Rankine-Hugoniot function, $R(X)$, is defined by

$$R(X) \equiv X^2 H(X) - \frac{\sin^2\theta_1}{M_{A1}^2} J(X) . \tag{23}$$

The hydrodynamic Rankine-Hugoniot function, when it is expressed in terms of the variable X, becomes

$$H(X) = \left(\frac{\gamma + 1}{\gamma - 1}\right) X + \frac{V_C^2 - V_{X1}^2}{V_{x1}^2} , \tag{24}$$

where V_c, given by

$$V_C^2 = \frac{\gamma + 1}{\gamma - 1} C_{11}^2 - \frac{2C_{S1}^2}{\gamma - 1} , \tag{25a}$$

will define the upper limit speed for switch-off shocks. The function $J(X)$ is defined by

$$J(X) \equiv \frac{\gamma}{\gamma - 1} X^2 - X \left(Y - \frac{1}{\gamma - 1}\right) + \frac{Y(1 - Y)}{\gamma - 1} . \tag{25b}$$

Equation (25), a quartic, has four solutions, one of which is trivial; solving $R(X) = 0$ then determines the other three solutions. The sign of the variable X measures whether V_x is above or below C_I. Once r or X is determined from $R(X) = 0$, the remaining physical quantities may be determined by algebraic substitution; the downstream transverse field is determined by (21a), and

$$\frac{V_y}{V_{x1}} = \frac{(Y - X) \sin\theta_1 \cos\theta_1}{X M_{A1}^2} , \tag{26a}$$

$$\frac{T}{T_1} = \frac{rP}{P1} = \frac{C_s^2}{C_{s1}^2} = r\left\{1 + \gamma M_{s1}^2 (Y-X) + \frac{\gamma}{2} \frac{\sin^2\theta_1 C_{A1}^2}{C_{s1}^2}\left(1 - \frac{Y^2}{X^2}\right)\right\} . \tag{26b}$$

4.2.   De Hoffmann - Teller frame

De Hoffmann and Teller[1] pointed out that it is sometimes convenient to transform the coplanar MHD Rankine-Hugoniot relations to a frame of reference in which the electric field is zero. To transform the electric field in the standard normal incidence frame, $E_z = - V_{x1}B_{y1}/c$, to zero, we must go to the frame of reference moving with the velocity

$$V_{HT} = \hat{e}_y \frac{V_{x1}B_{x1}}{B_x} , \tag{27}$$

with respect to the standard normal incidence frame. In the so-called De Hoffmann-Teller frame, the upstream flow velocity,

$$V_1 = \left(V_{x1}, \frac{V_{x1}B_{x1}}{B_x} \right) , \tag{28}$$

is parallel to the magnetic field. Since the transformed upstream electric field $E = 0$, E is also zero downstream, and the fluid velocity is parallel to the magnetic field downstream, as well,

$$V_2 = \left(V_{x2}, \frac{V_{x2}B_{x2}}{B_x} \right) . \tag{29}$$

The coplanar jump relations reduce to a form that is easy to remember in the De Hoffmann-Teller frame:

$$[\rho V_x]^* = 0 , \tag{30a}$$

$$\left[\rho V_x^2 + P + \frac{B_x^2}{8\pi} \tan^2\theta \right]^* = 0 , \tag{30b}$$

$$[\rho \tan\theta \, (V_x^2 - C_I^2)]^* = 0 , \tag{30c}$$

$$\left[\frac{\rho V_x^3}{2} (1 + \tan^2\theta) + \frac{\theta V_x P}{\gamma - 1} \right]^* , \tag{30d}$$

where $\tan\theta = B_y/B_x$. Since equation (1f) is automatically satisfied in the De Hoffmann-Teller frame, algebraic manipulations are somewhat simplified. However, equations (30) require that $B_x \neq 0$, and break down for perpendicular shocks. Indeed they fail whenever B_x is sufficiently small that V_{HT} approaches the speed of light. We choose to work with the MHD Rankine-Hugoniot relations in the normal incidence frame, because of their greater generality.

5.   CLASSIFICATION OF MHD SHOCKS

5.1.   Definition of Rankine-Hugoniot States

The Rankine-Hugoniot relations find and relate homogeneous and stationary flow states of local thermal equilibrium whose mass, momentum, and energy fluxes, and transverse electric fields are the same. The MHD shocks are transitions between pairs of stationary states whose order is determined by the requirement that the specific entropy downstream exceed that upstream. In this section, we classify the Rankine-Hugoniot states according to the relationship between the flow speed and the small amplitude wave speeds in each state. In the next section, we ascertain the ordering of their entropy densities and enumerate the possible shocks.

Figure 1 shows the curves $F(r,b) = 0$ and $Z(r,b) = 0$ for a choice of flow parameters for which there exist four Rankine-Hugoniot states. The upper supersonic and lower subsonic branches of $F = 0$ are drawn with heavy solid lines, and the two $Z = 0$ hyperbolae are drawn with dot-dash lines. The regions where Z is positive are shaded. The dashed horizontal line is the asymptote of the $Z = 0$ hyperbolae on which $X = 0$. Since the magnitude of the local flow speed, $|V_x|$ equals the intermediate speed on $X = 0$, $|V_x|$ exceeds the local intermediate speed on the upper hyperbola, and is less than the intermediate speed on the lower hyperbola.

The curves $F(r,b) = 0$ and $Z(r,b) = 0$ intersect at four points, which we label 1,2,3,4 in order of increasing density or decreasing velocity contrast, r. We chose to define the "upstream" parameters listed in Figure 1 in the standard normal incidence frame at point 1, so that the Rankine-Hugoniot solution for point 1 is the trivial one, $r = 1$, $b_1 = \sin\theta_1$. However, since points 1,2,3, and 4 are equivalent under the conservation laws, the "upstream" point labelled by subscript 1, and defined by the parameters M_{A1}, M_{S1}, and θ_1, could actually refer to any of the Rankine-Hugoniot states.

We have already seen that the flow speed $|V_x|$ exceeds the intermediate speed at points 1 and 2, and is less than the intermediate speed at points 3 and 4. Its relationship to the fast and slow speeds may be ascertained by the following argument. The curves $F = 0$ and $Z = 0$ must intersect for a Rankine-Hugoniot solution to exist. Figure 1 indicates that the gradients of r with respect to b along $F = 0$ and $Z = 0$ satisfy $(dr/db)_F > (dr/db)_Z$ at points 1 and 3, and the reverse inequality at points 2 and 4. After computing these gradients in the standard normal incidence frame at point 1,

$$(dr/db)_F = -\sin\theta_1/M_{A1}^2(1 - 1/M_{S1}^2) , \qquad (31a)$$

$$(dr/db)_Z = - Y/\sin\theta_1 . \tag{31b}$$

we may maneuver the inequality, $(dr/db)_F > (dr/db)_Z$, into the form

$$D_{FS}[V_x] = [V_x^2 - C_F^2][V_x^2 - C_{SL}^2] > 0 , \tag{32}$$

where C_F and C_{SL} are the fast and slow MHD speeds at point 1. Since the flow speed exceeds the intermediate speed, which in turn is larger than the slow speed, (32) implies that $|V_x|$ is above the fast speed at point 1.

We can extend the same reasoning to the three remaining Rankine-Hugoniot states, since their gradients, $(dr/db)_F$ and $(dr/db)_Z$, have the forms (31a,b) when local values of the Mach numbers and propagation angle are used. Then, because the downstream flow speed at point 2 still exceeds the intermediate speed, the reversal of the inequality (32) implies that $|V_x|$ is below the fast speed. The flow speed is less than the intermediate speed but D_{FS} is positive at point 3, so that $|V_x|$ exceeds the slow speed. Since the sign of D_{FS} reverses at point 4, the flow speed has to be less than C_{SL} there.

In summary, the four solutions to the coplanar Rankine-Hugoniot relations may be classified by the relationships of $|V_x|$ to the local MHD characteristic speeds as follows:

$$V_x^2 > C_F^2 > C_I^2 \text{ at point 1}$$

$$C_F^2 > V_x^2 > C_I^2 \text{ at point 2}$$

$$C_I^2 > V_x^2 > C_{SL}^2 \text{ at point 3}$$

$$C_I^2 > C_{SL}^2 > V_x^2 \text{ at point 4}$$

Several special cases are not illustrated by Figure 1. For example, if the upstream speed equals the fast (or slow) speed, the upper (or lower) $Z = 0$ hyperbola is tangent to the curve $F = 0$, points 1 and 2 (or points 3 and 4) coalesce, and we retrieve small amplitude fast (or slow) discontinuities. When $|V_x|$ equals the intermediate speed at points 2 or 3, we obtain the switch-on and switch-off shocks to be defined shortly.

5.2.   Thermodynamic causality, and the definitions of fast, intermediate and slow shocks

The conservation laws expressed by F(r,b) = 0 and Z(r,b) = 0 allow twelve different transitions, from, say, state 1 to state 2, state 2 to state 1, 1 to 3, 3 to 2, and so on, where our convention is to list the upstream state first. Not all the possible combinations of Rankine-Hugoniot states will be realized as shocks. To decide which ones can actually occur, one has to impose additional conditions derived from physical reasoning beyond that which went into the formulation of the conservation laws. Such "admissibility" conditions express the fact that shocks ought to develop causally in a time-dependent flow field. We will take up the so-called MHD evolutionary conditions shortly. Here, we discuss the most elementary admissibility condition, thermodynamic causality.

The MHD Rankine-Hugoniot conditions are symmetric with respect to an exchange of the upstream and downstream states because they do not take dissipation into account. However, thermodynamic causality requires that the shocks be irreversible. The specific entropy downstream must exceed that upstream. One can use the Navier-Stokes model of dissipation to show that the entropy density, S, in the Rankine-Hugoniot states is ordered as follows

$$S\ (4) \geq S\ (3) \geq S\ (2) \geq S\ (1)\ ,$$

from which it follows that $T(4) \geq T(3) \geq T(2) \geq T(1)$, where T is the temperature. Since this result can also be proven internally from the Rankine-Hugoniot relations themselves, at the cost of burdensome algebra,[2,3] the ordering of the entropy density in the Rankine-Hugoniot states does not depend upon the nature of the dissipation, so long as it returns the system to local thermodynamic equilibrium. The entropy conditions above reduce the number of allowed transitions from twelve to six, of types 1-2, 1-3, 1-4, 2-3, 2-4, and 3-4. Since $r(1) \geq r(2) \geq r(3) \geq r(4)$, the admissible MHD shocks are all compressional: the downstream density is larger than that upstream.

The six MHD shocks may be divided into two groups according to the direction of the transverse magnetic field downstream, which is the same on either side of the 1-2 and 3-4 shocks, and opposite for the other four, as is obvious in Figure 1. We call the 1-2 and 3-4 shocks fast and slow , respectively, because they reduce to the fast and slow waves in the small amplitude limit. While both are compressional, the magnetic field strength increases across the fast shock and decreases across the slow shock. The normal component of the shock frame flow velocity exceeds the intermediate speed on both sides of oblique fast shocks, and it is smaller than the intermediate speed on both sides of slow shocks. The remaining four shocks are called intermediate. The direction of the transverse magnetic field rotates by 180 degrees and the flow speed goes from above to below the intermediate speed across all four of them.

5.3.   Switch-on and Switch-off Shocks

The definitions of the fast, intermediate, and slow MHD shocks introduced in the previous section are unambiguous for oblique propagation, when the transverse magnetic field is nonzero and the coplanarity plane is defined on both sides of the shock. In this section, we discuss the special cases that arise when the coplanarity plane cannot be defined on at least one side of the shock. In these cases, the azimuthal symmetry of the one-dimensional MHD equations for parallel propagation comes into play.

When the transverse magnetic field is zero in at least one of the Rankine-Hugoniot states, we must return to the general relations for non-coplanar propagating discontinuities derived in Section 2.3, $F(r, \mathbf{b_T}) = 0$, and $Z(r, \mathbf{b_T}) = 0$. When $\mathbf{b_{T1}}$ is zero, F depends upon b_T^2 and not upon the direction of $\mathbf{b_T}$, and the surface $F(r, \mathbf{b_T}) = 0$ is an azimuthally symmetric figure of rotation about the r-axis in the three-dimensional $(r, \mathbf{b_T})$ space. Furthermore, the equation $Z = 0$ reduces to

$$Z(r, \mathbf{b_T}) = \mathbf{b_T}(r - 1/M_{A1}^2) = \mathbf{b_T}X = 0 \qquad (33)$$

(Note that the intermediate Mach number, M_{I1}, and the Alfven Mach number, M_{A1}, are identical because of the parallel-Alfven degeneracy.) Equation (33) is satisfied only along the r-axis, where $\mathbf{b_T} = 0$, and on the plane defined by $X = 0$ where the local flow speed equals the intermediate speed. Thus the $Z = 0$ hyperbolae obtained for oblique propagation degenerate into their asymptotes in the limit of parallel propagation. Because of azimuthal symmetry, the $Z = 0$ solution generates a line and a plane.

The Rankine-Hugoniot states are intersections of the $F(r, \mathbf{b_T}) = 0$ surface with the r-axis and with the $X = 0$, intermediate speed, plane. Depending upon parameters, the $X = 0$ plane either intersects the $F(r, \mathbf{b_T}) = 0$ surface or it does not. If it does not, there will be only two Rankine-Hugoniot states, those where the $F = 0$ surface intersects the r-axis. In this case, the only possible shock is a purely hydrodynamic one, since the upper and lower branches of $F = 0$ satisfy $H(r) = 0$ for the same magnetic field.

Figure 2 shows a two-dimensional planar cut through the r-axis for a case where the intermediate speed plane $X = 0$ does intersect the $F = 0$ surface. This intersection generates a dense set of Rankine-Hugoniot points forming a circle in the X = 0 plane, which we call the 2-3 circle. The intersection of the r-axis with $F = 0$ generates isolated points of types 1 and 4 related by $H(r) = 0$. The fast switch-on shock is a transition between point 1 and any point on the 2-3 circle. It is given its name because it has transverse components of magnetic field and fluid velocity downstream though none exist upstream. The rotational discontinuity is a transition between any two points on the 2-3 circle, and the slow switch-off shock is a transition between any point on the 2-3 circle and point 4. Here, the transverse component of magnetic field upstream is switched off downstream. The normal component of the shock frame flow velocity equals the intermediate speed downstream of the switch-on shock, upstream of the switch-off shock, and upstream and downstream of the rotational discontinuity.

The coplanar Rankine-Hugoniot relations, (22), reduce to $X^2H(X) = 0$ when $\sin^2\theta_1 = 0$. Because of azimuthal symmetry, this solution is valid for any orientation of the transverse magnetic field. When the upstream state is of type 1, the solution $H(X) =$

0 describes a hydrodynamic shock with $\mathbf{b}_T = \mathbf{U}_T = 0$ downstream. The degenerate pair of solutions to $X^2 = 0$ represent the switch-on shock. The magnitude of the transverse field on the 2-3 circle may be found by solving $F = 0$ and $X = 0$ simultaneously:

$$\mathbf{B}_T/B_x = \frac{\left\{ (\gamma - 1)(V_{x1}^2 - C_{A1}^2)\,(V_{C1}^2 - V_{x1}^2) \right\}^{1/2}}{C_{A1}^2}, \tag{34a}$$

where

$$V_{C1}^2 = \frac{\gamma + 1}{\gamma - 1}\, C_{A1}^2 - \frac{2}{\gamma - 1}\, C_{S1}^2 = C_{A1}^2 + \frac{2\left(C_{A1}^2 - C_{S1}^2\right)}{\gamma - 1}. \tag{34b}$$

Since $\mathbf{B}_T$ is real only for $C_{A1}^2 \le V_{x1}^2 \le V_{C1}^2$, switch-on shocks exist only for Alfven Mach numbers between unity and the upper limit, V_{C1}/C_{A1}, which itself is limited to the range

$$1 \le V_{C1}^2/C_{A1}^2 \le \frac{(\gamma + 1)}{(\gamma - 1)}, \tag{35}$$

depending upon the sound speed. Its largest value is achieved for $C_{S1} = 0$, and its smallest value, $V_{C1} = C_{A1}$, occurs when $C_{S1} = C_{A1}$. It follows that switch-on shocks cannot exist when the sound speed exceeds the Alfven speed upstream.

It is straightforward to show that

$$V_T/V_{x1} = (1/M_{A1}^2)(\mathbf{B}_T/B_x) \tag{36a}$$

on the 2-3 circle, which implies that

$$V = B/(4\pi\rho)^{1/2}, \tag{36b}$$

where $V = (V_x, V_T)$, and $B = (B_x, \mathbf{B}_T)$. Thus, the flow velocity is parallel to the magnetic field downstream of the switch-on shock, consistent with the fact that $E = -(V \times B)/c$, which is zero by assumption at point 1, must also be zero at any downstream point on the 2-3 circle. The electric field is obviously zero upstream and downstream of the hydrodynamic shock as well. Finally, we may show that

$$T/T_1 = \left(1/M_{A1}^2\right)\left[1 + \gamma M_{S1}^2\, Y - b_T^2/\beta_1\right], \tag{37}$$

on the 2-3 circle, where we define β to be the ratio of the fluid to the magnetic pressure, $\beta_1 = 2C_{S1}^2/\gamma C_{A1}^2$.

The properties of the switch-off shock could be found by solving $H(r) = 0$ to determine point 4, and then relating the flow properties on the 2-3 circle to those at point 4. However, we may also solve the Rankine-Hugoniot relations in the standard normal incidence frame of the slow shock itself by choosing the fiducial point $r = 1$ to be a point of type 3, with $U_{T1} = 0$. By substituting the condition for a switch-off shock, $M_{11}^2 = 1$, into the algebraic form of the Rankine-Hugoniot relations, (22, 23), we find two roots, t_+ and t_-, where $t = (r - 1)$ and

$$(\gamma + 1)t_\pm = - \left[1 - 1/M_{S1}^2 - (\gamma/2)\tan^2\theta_1 \right]$$

$$\pm \left\{ \left[1 - 1/M_{S1}^2 - (\gamma/2)\tan^2\theta_1 \right]^2 + (\gamma + 1)\,\tan^2\theta_1 \right\}^{1/2} \qquad (38)$$

The root t_+ corresponds to the (inaccessible) type 1 Rankine-Hugoniot point, while t_- defines the normal component of the flow speed in the type 4 state downstream of the switch-off shock. The transverse velocity at point 4 is (minus) the Alfven speed based on the upstream transverse field and the downstream density,

$$V_T = - B_{T1}/(4\pi\rho)^{1/2} \qquad (39)$$

in the standard normal incidence frame of the switch-off shock. This result is consistent with (36b) which was obtained in the normal incidence frame at point 1.

Finally, switch-on and switch-off shocks differ from oblique shocks in one interesting aspect. It is conventional to define a shock by specifying its upstream state and speed, which is sufficient to determine the state downstream of oblique shocks. Alternatively, we could specify their downstream states and speeds, and the Rankine-Hugoniot relations would fix their upstream states. We may carry out the computation in either direction. Not so for the switch-on and switch-off shocks. The upstream state and shock speed do not determine the azimuthal orientations of the transverse magnetic field and flow velocity downstream of the switch-on shock, and the downstream state and shock speed do not fix the orientations of the transverse field and velocity upstream of the slow shock. Of course, if we are given both the upstream and downstream states, we may always calculate the speed of the shock that connects them.

5.4.   Parallel MHD Shocks

Here, we clarify the relationships between the definitions of fast, intermediate and slow shocks, and the hydrodynamic and switch-on shocks obtained for parallel propagation. We will point out the various possibilities as we vary the Alfven Mach number and the ratio of the sound and Alfven speeds.

The two panels of Figure 3 plot the curves $F(r, b_y) = 0$ (solid line) and $Z(r, b_y) = 0$ (dashed line) for cases in which $M_{A1} = M_{I1}$ exceeds unity and $C_{S1} = 0$ upstream. Because of azimuthal symmetry, these curves are the same for any orientation of b_T. When $\gamma = 5/3$, the switch-on upper limit Mach number, which is given by $M_{A1}^2 = (\gamma + 1)/(\gamma - 1)$ when $C_{S1} = 0$, is 2. Several $F = 0$ and $Z = 0$ curves are displayed in the right-hand panel of Figure 3 for values of M_{A1} both above and below 2. The $X = 0$ plane cuts the $F = 0$ surfaces only for Alfven Mach numbers less than 2. All the $F = 0$ surfaces cross the r-axis at the same two points, which satisfy the hydrodynamic Rankine-Hugoniot relation for a strong shock, and are located at $r = 1$ and $r = (\gamma - 1)/(\gamma + 1) = 1/4$.

When $M_{A1} < 2$, we obtain switch-on shocks and a hydrodynamic shock, as we indicated earlier. According to the definitions introduced in Section 5.2, the hydrodynamic shock is a 1-4 intermediate shock. It is an intermediate shock because the flow speed goes from above to below the intermediate speed. Since, in addition, the downstream flow speed is less than the sound speed, which is the MHD slow speed for parallel propagation, the downstream state is of type 4. When the Alfven Mach number at point 1 satisfies $M_{A1}^2 = (\gamma + 1)/(\gamma - 1)$, the switch-on solution, $r = 1/M_{A1}^2$, and the hydrodynamic solution, $r = (\gamma - 1)/(\gamma + 1)$, are identical. When $M_{A1}^2 > (\gamma + 1)/(\gamma - 1)$, the $X = 0$ plane does not intersect the $F = 0$ surface, and all that remains is the hydrodynamic shock, which define to be a 1-2 fast shock, since the flow speed exceeds the intermediate speed downstream.

The left-hand panels of Figure 3 display the $F(r,b_y) = 0$ and $Z(r,b_y) = 0$ curves for a fixed Alfven Mach number, 1.45, and several values of C_{S1}/C_{A1}. When the shock speed is below the switch-on critical speed, we again obtain a switch-on shock and a 1-4 hydrodynamic shock. As C_{S1}/C_{A1} increases, V_{C1} decreases until V_{x1} eventually exceeds the critical speed, and there is only the 1-2 hydrodynamic shock.

The case when $C_{S1} \leq |V_{x1}| \leq C_{A1}$ upstream is not encompassed by Figure 3. Here, the $X = 0$ plane is above the $F = 0$ surface, and we obtain only a 3-4 slow shock.

In summary, we obtain switch-on and switch-off shocks only when the $X = 0$ plane intersects the $F = 0$ surface. Otherwise, there are only two Rankine-Hugoniot states, which are those upstream and downstream of a hydrodynamic shock.

5.5.   Relationship between switch-on shocks, the rotational discontinuity, and intermediate shocks

The magnetohydrodynamic Rankine-Hugoniot relations for oblique coplanar shocks have four solutions, three from the cubic $R(X) = 0$, and one, $r = 1$, that denotes the point at which the sonic and Alfven Mach numbers, and the propagation angle, are defined. The four solutions are not always all real. The cubic has either one or three real solutions, so that the Rankine-Hugoniot relations have either two or four real solutions. A pair of solutions turns complex because one of the two branches of $Z(r,b_y) = 0$ fails to intersect $F(r,b_y) = 0$. Since the upper and lower branches of $Z = 0$ are responsible for intersections of types 1 and 2 and types 3 and 4, respectively, we conclude that when the Rankine-Hugoniot relations have only two solutions, they will be either of types 1 and 2, or of types 3 and 4, and there will be only one type of shock, either fast or slow. This situation prevails for most choices of flow parameters. Intermediate shocks exist <u>only</u> when the Rankine-Hugoniot relations have four <u>real</u> solutions.

Equating the discriminant of the cubic $R(X) = 0$ to zero determines when two of the Rankine-Hugoniot solutions become equal, just before they turn complex. The dashed line in the left-hand panel of Figure 4 shows, for $\beta_1 = 0.1$, where the discriminant of the Rankine-Hugoniot cubic is zero in a Friedrichs diagram format. The flow parameters are defined in the normal incidence frame of the upstream state, $r = 1$, which may be a point of type 1, 2, or 3. The upstream flow speed is the radius vector and the shock propagation angle, θ_1, is the polar angle in polar coordinates. The speeds of the small amplitude fast, intermediate, and slow waves are displayed for reference. The upstream flow speeds for which there are four real Rankine-Hugoniot solutions are shaded, lightly for upstream states of type 1, and heavily for those of type 2. The dark solid line shows the speeds of switch-on shocks and the rotational discontinuity.

The four panels to the right in Figure 4 display the curves $F(r,b_y) = 0$ (solid lines) and $Z(r,b_y) = 0$ (dotted lines) that correspond to the upstream points labelled "a", "b", "c", and "d" in the Friedrichs diagram to the left. The flow speed at point a is below the intermediate speed and above the slow speed; only a 3-4 slow shock is possible since the upper $Z = 0$ hyperbola does not intersect the $F = 0$ curve. The type 2 upstream point b is in the region of four real Rankine-Hugoniot solutions so that 2-3 and 2-4 intermediate shocks and a 3-4 slow shock can exist. (The upper $Z = 0$ hyperbola also intersects $F = 0$ at a point of type 1 which is inaccessible from point 2.) The type 1 upstream point c generates 1-2, 1-3, and 1-4 shocks because it too lies in the region of four real solutions. For the type 1 upstream point d, the lower $Z = 0$ hyperbola does not intersect the $F = 0$ curve and there is only a 1-2 fast shock.

Figure 5 is a Friedrichs diagram for β unity in the upstream state. Both upstream and downstream states in the region of four real Rankine-Hugoniot solutions are shown, in contrast with Figure 4. Since the switch-on critical Mach number is small when β is unity, the region of type 1 intermediate shock states is also small. Solid shading locates the type 2 states available to intermediate shocks. The hatched and dotted regions show the types 3 and 4 states available as downstream states for intermediate shocks.

Intermediate shocks are clearly related to the switch-on shock and rotational discontinuity. Their speeds are between the intermediate speed and an upper limit which reduces to the switch-on critical speed for parallel propagation. The orientations of the magnetic field and fluid velocity downstream of the switch-on shock and rotational discontinuity are arbitrary, and the intermediate shocks emerge when the azimuthal symmetry of parallel MHD is broken by the introduction of a small transverse magnetic field. For example, consider Figure 6, which shows the dependence of the intermediate shock upper limit speed upon the ratio of specific heats, γ, for $C_{S1} = 0$. The switch-on critical Mach number, $M_{A1}^2 = (\gamma + 1)/(\gamma - 1)$, increases as γ decreases. Since intermediate shocks may be thought of as switch-on shocks with broken azimuthal symmetry, the upper limit speed for all intermediate shocks also increases.

In summary, intermediate shocks express the breakdown of hyperbolic MHD. It seems useful to divide one-dimensional MHD conceptually into hyperbolic and nonhyperbolic regimes, according to whether the conservation laws permit 2 or 4 stationary points. In the hyperbolic regime, there can be only one kind of shock, either fast or slow, and intuition drawn from fluid mechanics is pertinent. In the non-hyperbolic regime, it will be necessary to grapple with the connectivity and topology of the magnetic field, and intermediate shocks are possible.

6.   THE MHD EVOLUTIONARY CONDITIONS

One can imagine initializing a numerical integration of the MHD equations with the various MHD shocks and finding that they continue to exist, so long as the boundary conditions do not change and the shocks are stable. However, the boundary conditions change in real physical systems, and we are interested only in those shocks that form naturally and can adjust to variable boundary conditions. We now address this issue of dynamical causality.

Before we discuss the MHD evolutionary conditions formally, we introduce an intuitive evolutionary argument about fast shocks. We showed earlier, by examination of the properties of the curves, $F(r,b) = 0$ and $Z(r,b) = 0$, that the flow speed exceeds the fast speed upstream and is below the fast speed downstream of fast shocks. We now argue that these conditions are necessary to the dynamical stability of the fast shock. Let us imagine that a piston far downstream launches a steepening simple wave in the fast mode. We divide the steepening wave into a succession of small amplitude step waves. A fast shock starts to form when one of the small amplitude waves overtakes another. Thereafter, waves which overtake the newly formed shock will strengthen it, until dissipation eventually balances steepening to create a steady shock profile. The piston can influence the subsequent evolution of the shock only if downstream fast waves continue to overtake the shock. Thus, the fast speed must be larger than the flow speed downstream. On the other hand, if the fast speed exceeded the flow speed upstream, fast waves could run ahead of the shock from downstream, and the shock profile would be unsteady. Therefore, the fast speed must be less than the flow speed upstream. These conditions are necessary for the piston to create a shock and to influence its subsequent strength until it reaches a steady state.

The above argument, which is due to Kantrowitz and Petschek[4], elucidates the physics implied by the theory of MHD characteristics. What does the formal theory say? According to Lax's[5] treatment of n-dimensional strictly hyperbolic systems, the number of characteristics leaving the shock, either upstream or downstream, must be one less than the number of independent variables, i.e., n - 1, for the shock to be evolutionary. Lax drew his conclusion from the theory from the theory of free boundary value problems, where the number of relations to be imposed on the two sides of a discontinuity must equal the number of characteristics impinging on the discontinuity from either side in order for the solution to be well posed. Since after specification of the shock speed, the Rankine-Hugoniot conditions relate n - 1 of the 2n quantities defining the upstream and downstream states, there must be n + 1 characteristics entering the shock and n - 1 leaving it.

It is easy to see that an evolutionary shock should take the normal component of the shock frame flow velocity from above to below only one characteristic speed, the conclusion motivated by Kantrowitz and Petschek's physical argument. Another way of phrasing this conclusion is to say that two characteristics of the same family end on a shock that is evolutionary.

For magnetohydrodynamics, $n = 7$, and so six characteristics must leave and eight characteristics must enter an evolutionary shock. Ascertaining which of the discontinuities predicted by the MHD Rankine-Hugoniot relations are evolutionary amounts to a straightforward counting of characteristics, which is summarized in Figure 7. The horizontal and vertical axes are the upstream and downstream flow speeds, V_{x1} and V_{x2}, respectively. Each axis is divided into regions by the MHD characteristic speeds, and each region corresponds to one Rankine-Hugoniot state. For example, upstream and downstream states of type 2 are in the vertical column defined by $C_{I1} \leq V_{x1} \leq C_{F1}$ and the horizontal row defined by $C_{I2} \leq V_{x2} \leq C_{F2}$, respectively. A pair of Rankine-Hugoniot states is then the box formed by the intersection of a row with a column. The number of characteristics leaving the discontinuities corresponding to each of the sixteen possible pairs of Rankine-Hugoniot states is shown in each box. Those where $n = 6$ are shaded. The discontinuities which survive this evolutionary test are the 1-2 fast shock, the 2-3 intermediate shock, and the 3-4 slow shock. The 1-3, 1-4, and 2-4 intermediate shocks are not evolutionary because they take the flow speed from above to below more than one characteristic speed.

Akhiezer, Liubarskii, and Polovin[6] argued that only the evolutionary discontinuities are stable. They postulated that it is possible to form thin discontinuities with upstream and downstream flow speeds corresponding to each box in Figure 7. They then linearized the hyperbolic conservation laws about the upstream and downstream states, and generated the general form of the perturbed solution using Laplace transform techniques. Perturbations about nonevolutionary discontinuities exponentiated without bound, either far upstream, or far downstream, or both.

Jeffrey and Taniuti[7] argued that the 2-3 intermediate shock should also be considered nonevolutionary, because the information about changes in B_z carried by the intermediate characteristic family is orthogonal to the information about B_y needed to specify the states bounding the intermediate shock. Mathematically, this is expressed by the fact that the characteristic matrix A defined in Chapter 3 splits into two square submatrices, a 5x5 matrix defining the fast, slow, and entropy modes, and a 2x2 matrix for the intermediate mode. Consequently, the boundary conditions specifying the discontinuities contained in the two submatrices should be applied separately. It seems the small amplitude intermediate wave does not carry the right information to make an intermediate shock. Furthermore, the intermediate simple wave does not steepen, so how can a thin intermediate shock form?

Kantrowitz and Petschek[4] generated two further arguments against the existence of intermediate shocks: the intermediate shock is an isolated solution of the MHD Rankine-Hugoniot relations that has no neighboring solution corresponding to a small change in the external boundary conditions, and the intermediate shocks are extraneous, meaning that any boundary conditions calling for intermediate shocks can always be satisfied using the rotational discontinuity. We take up these two arguments in order.

All intermediate shocks have the properties that B_y changes sign across the shock, and that the flow speed exceeds the intermediate speed upstream and is less than the intermediate speed downstream. Let us suppose that a conducting piston has created a coplanar intermediate shock somewhere in a flow field. Since B_z and V_z are zero downstream, the piston motion required to generate the shock was in the x-y plane. Now, imagine that the piston subsequently moves in the z-direction, and therefore launches an intermediate wave. There is no place in the flow field where the intermediate wave can stand, since it would catch the shock from behind, and be blown into it from ahead. Thus a flow containing an intermediate shock does not have any neighboring steady shock solutions corresponding to a small change in the piston boundary conditions that rotates the direction of the magnetic field by an arbitrarily small amount. This argument led Kantrowitz and Petschek to conclude that intermediate shocks are not structurally stable and are physically unrealizable.

Kantrowitz and Petschek argued, furthermore, that any piston boundary conditions that require B_y to change sign can always be satisfied by inserting a 180 degree rotational discontinuity into a flow field containing evolutionary fast and slow shocks and/or rarefactions. Therefore, intermediate shocks are not needed, and should be considered "extraneous" solutions of the Rankine-Hugoniot relations. They applied a similar argument to the nonevolutionary, parallel propagating 1-4 hydrodynamic shock discussed in Section 5.4. How are we to account for situations in which a non-conducting piston produces a purely compressional sound wave that will steepen to form a shock whose downstream speed is less than the intermediate speed? The switch-off shock provides an answer. All the Rankine-Hugoniot points in Figure 2 have the same mass, momentum, and energy fluxes, and their electric fields measured in the normal incidence frame at point1 all transform to zero. Thus, a switch-on shock can be followed by a switch-off shock which travels at the speed of the fluid downstream of the switch-on shock. If the piston created both at the same time, the composite system of "attached" shocks would be indistinguishable, except possibly for its interior structure, from the 1-4 hydrodynamical shock.

Arguments such as those summarized above suppressed all consideration of intermediate shocks for the past quarter century. This conventional wisdom has recently been challenged by numerical studies of the time-dependent, one-dimensional MHD Navier-Stokes equations, which reveal that stable intermediate shocks can evolve from smooth boundary conditions.[8-14] Consequently, we will not suppress discussion of nonlinear intermediate waves and shocks in this book. Nonetheless, the reader should not ignore the above arguments against intermediate shocks. However, as our exposition unfolds, he should bear in mind the following points :

*The evolutionary conditions stem from the theory of strictly hyperbolic systems, yet the parallel-Alfven degeneracy is an essential feature of MHD.

*There are circumstances in which intermediate waves do steepen.

*The MHD Rankine-Hugoniot relations convey a misleading subliminal message. Because ideal MHD is scale-free, it permits finite amplitude discontinuities. It is therefore tempting to presume that all pairs of Rankine-Hugoniot states will always be separated by a discontinuity in the limit that the dissipation rate is small. However, the Rankine-Hugoniot conditions only relate states of local thermodynamic equilibrium.

They do not guarantee that there exists a unique, steady structure solution that converges to a thin discontinuity in the weak dissipation limit.

*Akheizer, et al's stability analysis, which concluded that only evolutionary shocks are stable, was based on the assumptions that MHD is hyperbolic, and that all its Rankine-Hugoniot transitions are thin.

*The argument that a 180 degree rotational discontinuity can always substitute for an intermediate shock may be faulty, since the rotational discontinuity is neither steady nor thin, when dissipation is taken into account.

*Like intermediate shocks, the 1-2 fast shock also has no neighboring steady solution corresponding to a slight deviation from coplanarity, yet no one argues about *its* existence.

*The fact that integration of the MHD Navier-Stokes equations gives a different result than reasoning based upon ideal MHD suggests that the questions of evolutionarity and structure cannot be separated in the case of intermediate shocks.

7.   CONCLUDING REMARKS

The MHD Rankine-Hugoniot conditions relate uniform, stationary, dissipation free states of local thermodynamic equilibrium asymptotically far upstream and downstream of the shock. When particle collisions provide the shock dissipation, it is evident that full thermodynamic equilibrium can be achieved downstream. In collisionless plasma, the shock dissipation will have to be provided by collective processes. Several such processes are expected to occur and to have different scalelengths. Some may be rather subtle. For example, MHD requires that the plasma pressure be isotropic upstream and downstream. The collective processes which isotropize the pressure can act on scalelengths much longer than those which account for most of the entropy production in the shock front. Similarly, the turbulent processes in the maximum dissipation layer of the shock can heat ions and electrons at different rates, and the scalelength over which the ion and electron temperatures equalize can be very long. It is important to remember that the Rankine-Hugoniot relations apply only to states separated by a distance greater than the longest scalelength of all the processes needed to achieve local thermodynamic equilibrium downstream. Only then are they independent of the nature of the processes that bring the shocked plasma to thermal equilibrium.

APPENDIX A. PROPERTIES OF MAGNETOHYDRODYNAMIC SHOCKS

Now that the MHD Rankine-Hugoniot relations can be solved with the aid of a small personal computer, finding the properties of the various shocks in different parameter regimes no longer presents the terrors it did when the Rankine-Hugoniot relations were first discussed in the literature. It is for this reason that our exposition above concentrated on issues of theoretical generality. Nonetheless, all who work experimentally on shock waves in plasmas are acutely aware of their parameter sensitivity. In this Appendix, we survey in broad outline the dependence of the properties of magnetohydrodynamic shocks upon their upstream parameters.

A.1.   Properties of Fast Shocks

We start by presenting analytical results that may be obtained in particularly simple limiting cases.

The fast shock is the only one whose speed is not limited, and is therefore the only one to which the results in the strong shock limit presented below apply. We already showed that when the magnetic field approaches zero, the MHD Rankine-Hugoniot relations approach the hydrodynamic Rankine-Hugoniot relations. Letting the Alfven Mach number approach infinity is formally equivalent to equating the magnetic field to zero. Thus, in the strong shock limit, the MHD fast shock approaches a strong hydrodynamic shock whose solution is

$$r = (\gamma - 1)/(\gamma + 1) , \tag{A.1}$$

which implies that the velocity contrast approaches a limit of 1/4 when $\gamma = 5/3$. We also find that the downstream transverse velocity, V_y, approaches zero as it does in hydrodynamic shocks. The downstream state is independent of the orientation of the upstream magnetic field in the strong shock limit. It is easy to see that the density and magnetic field compression ratios, ρ/ρ_1 and B_y/B_{y1}, approach the limit $(\gamma + 1)/(\gamma - 1) = 4$. These are the maximum compression ratios for fast shocks of any Mach number. Because they are finite, the downstream Alfven and intermediate speeds approach finite limits,

$$C_A^2 => \frac{\gamma - 1}{\gamma + 1} \, C_{11}^2 + \frac{\gamma + 1}{\gamma - 1} \, C_{A1}^2 \, , C_I^2 => \frac{\gamma - 1}{\gamma + 1} \, C_{11}^2 \, , \tag{A.2}$$

However, the sound speed increases without limit,

$$C_s^2 => \frac{2\gamma(\gamma - 1)}{\gamma + 1} \, V_{X1}^2 \, , \tag{A.3}$$

as the shock speed tends to infinity. This relation together with (A.1) implies that V_{x2}^2 approaches $C_s^2/5$ downstream and that the ratio of the internal energy flux downstream to the mechanical energy flux upstream,

$$\frac{2\gamma PV_x}{(\gamma - 1)\, \rho_1 V_{X1}^3} => \frac{4\gamma}{(\gamma + 1)^2} \tag{A.4}$$

approaches 15/16 when $\gamma = 5/3$. Thus, strong MHD shocks convert nearly all their energy into thermal energy, and invest only a small fraction in the magnetic field.

Perpendicular magnetosonic shocks are the other simple limiting case. It is easy to show that their downstream transverse velocity, V_y, is zero, and that their magnetic field and density compression ratios are equal. Magnetosonic shocks behave much like hydrodynamic shocks in which the sum of the fluid and magnetic pressures replaces the fluid pressure. When the ratio of specific heats, γ, is two, the analogy is exact, because then both the magnetic field and pressure respond two-dimensionally to density variations.

The properties of fast shocks depend critically upon the direction and strength of the upstream magnetic field, as well as upon their Mach number. Figure A.1 illustrates how fast shocks depend on the upstream field direction when the upstream plasma is as strongly magnetized as possible: when it is cold. The dependences upon the fast Mach number, M, of the density compression ratio (top panel), the magnetic field compression ratio (middle panel) and the ratio of the downstream internal energy density to the upstream flow energy density (bottom panel) are shown for shock normal angles, θ, ranging in fifteen degree increments between parallel (0°) and perpendicular (90°) propagation. (The Rankine-Hugoniot relations were solved in the standard normal incidence frame for this and all subsequent figures in this Appendix.) Parallel switch-on shocks exist between Mach numbers 1 and 2, and then give way to a strong hydrodynamic shock with $\rho_2/\rho_1 = 4$, when $\gamma = 5/3$. The more oblique the shock, the higher the Mach number at which it approaches the strong shock limit, but the strong shock limit will be a good approximation for all fast shocks with Mach numbers exceeding 5.

Note that the states downstream of shocks propagating at more than 45 degrees to the upstream field are similar, and differ considerably from their more parallel cousins. This is one reason why experimentalists distinguish between quasi-parallel ($\theta < 45°$) and quasi-perpendicular ($\theta > 45°$) shocks.

Figure A.2 illustrates how fast shocks depend upon their Mach number (M_{f1}), and upon the upstream field direction (θ_{NB1}) and strength (β_1). The top and bottom rows show cases with $\beta_1 = 0$ and 2 respectively. The first column displays contours of the ratio of the downstream to upstream total magnetic field against fast Mach number and shock normal angle; the second, of the velocity contrast (here: u_{2x}/u_1); and the third, of the ratio of fluid to magnetic pressure downstream, β_2. The difference between quasi-parallel and quasi-perpendicular shocks is evident in the magnetic field contours. The velocity contrast depends strongly on the shock normal angle for $\beta_1 = 0$, while it is virtually free of any dependence on θ_{NB1} when the fluid is weakly magnetized, $\beta_1 = 2$. Finally, since the ratio of fluid to magnetic pressure downstream, β_2, exceeds unity for Mach numbers exceeding 3 when the plasma upstream is cold, we infer that the fluid pressure exceeds the magnetic pressure downstream of all fast shocks with Mach numbers greater than 3.

Figure A.3 shows the dependence on C_{S1}^2/C_{A2}^2 of several properties of parallel fast shocks. Here, we define the fast Mach number, M_{F1}, to be V_{x1}/C_{A1} for $C_{S1} < C_{A1}$, and V_{x1}/C_{S1} for $C_{S1} > C_{A1}$, and contour the ratios of the downstream transverse magnetic field to the upstream total field (upper left panel), of the normal component of the downstream flow speed to the intermediate speed (upper right), and of the sound speed to the Alfven speed downstream (lower right), together with the denstiy compression ratio (lower left), against C_{S1}^2/C_{A2}^2 and M_{F1}. The switch-on shock region is where B_{y2} is nonzero and the downstream flow speed equals the intermediate speed. The 1-2 hydrodynamic shock is in the complement of the switch-on region; its properties are independent of the magnetic field, as is evident for $C_{S1}^2/C_{A2}^2 > 1$, where the normalization of the fast Mach number does not obscure this property.

A.2.   Properties of Intermediate Shocks

Figures A.4 and A.5 contour the velocity contrasts of 1-3 and 2-3 intermediate shocks, and 1-4 and 2-4 intermediate shocks, respectively, in a Friedrichs diagram format for an upstream $\beta = 0.1$. The panels to the left of each figure show the upper limit speed for intermediate shocks (dashed line) and the right-hand panels display the velocity contrasts. We have exaggerated the horizontal scales in the right-hand panels to emphasize the intermediate shock regions. The four intermediate shocks can be subdivided into two subclasses according to their downstream states, since the velocity contrast does not change in behavior as the upstream state changes from type 1 to type 2. The 1-3, 2-3 intermediate shocks are weak for intermediate Mach numbers near unity, and become progressively stronger as the shock speed approaches the upper limit speed. The 1-4, 2-4 intermediate shocks are never weak, and their velocity contrast lies within the relatively narrow range, $0.29 < r < 0.45$, for all propagation angles and speeds.

Figure A.6 contours the ratio of the downstream to upstream pressure for the 1-3, 1-4 (left) and 2-3, 2-4 (right) intermediate shock families. Once again, the 1-3, 2-3, shocks are weak when V_{x1} is close to the intermediate speed, while the 2-3, 2-4 shocks never have small pressure ratios. In either case, the pressure ratios are large for shock speeds near the upper limit speed. Thus, strong intermediate shocks can heat the plasma significantly.

Figures A.7 and A.8 contour the ratios of the downstream transverse velocity to the upstream normal velocity, and of the downstream transverse magnetic field to the upstream total field, respectively, for the 1-3, 2-3 (left panels) and the 1-4, 2-4 (right panels) intermediate shocks. The 1-3 shocks induce large transverse fields and velocity components downstream, while the 1-4 shock, which resembles a hydrodynamic shock, does not deflect the flow and magnetic field nearly as much. The 2-3 and 2-4 shocks are more nearly similar in this regard.

A.3.   Properties of Slow Shocks

Let us begin by contrasting the properties of quasi-parallel and quasi-perpendicular slow shocks. Parallel slow shocks exist only when the sound speed is less than the Alfven speed upstream, and their speeds lie in the range, $C_{S1} < V_{x1} < C_{A1}$. Parallel slow shocks are hydrodynamic. The MHD slow speed is zero for perpendicular propagation, and is very small for quasi-perpendicular propagation. Consequently, relatively strong slow shocks can move relatively slowly. In the limit of quasi-perpendicular propagation, in which we assume that $\tan^2\theta_1 \gg 1$, (38) has the approximate solution,

$$r = \frac{(\gamma - 1)/\gamma + \beta_1}{1 + \beta_1},$$

(A.5)

so that the velocity contrast, r, ranges between $(\gamma - 1)/\gamma$ when β_1 is small to unity when β_1 is large. Using the relation, $C_I^2 = rC_{I1}^2$, we may also show that

$$C_s^2 = C_{S1}^2 + \{(\gamma - 1)/\gamma\}\, C_{A1}^2$$

(A.6)

which implies that $\beta = 1 + \beta_1$, so that the downstream fluid pressure always exceeds the magnetic pressure. Thus, quasi-perpendicular slow shocks efficiently convert upstream magnetic energy into fluid pressure downstream. In addition, the fluid is accelerated in the transverse direction to a speed comparable with the Alfven speed given by (39).

The fact that the speeds of slow shocks are restricted to the range, $C_{SL1} < V_{x1} < C_{I1}$, implies that all slow shocks are weak when $C_{S1} > C_{A1}$; since the slow and intermediate speeds are very nearly equal in this parameter regime, the slow Mach number is always close to unity.

Figure A.9 contours the density compression ratio (here: n_2/n_1) and total magnetic field compression ratio against the slow Mach number, M_{SL}, and shock normal angle (θ_{NB1}) for $C_{S1}^2/C_{A1}^2 = 1/16$ (left column), 1/2 (middle), and 2 (right). The switch-on Mach number, $M_{SL} = C_{I1}/C_{SL1}$, is indicated by a dashed line in each panel. The upper left panel shows a division into distinct quasi-parallel and quasi-perpendicular regimes near the switch-on limit. The quasi-parallel shocks are essentially hydrodynamic in nature, while the quasi-perpendicular shocks are strongly magnetized and are described by (A.5) and (A.6) above. In the quasi-parallel regime, the magnetic field decompression ratio is virtually independent of the slow Mach number, while the decompression is strong and Mach number-dependent for quasi-perpendicular shocks. The slow shocks become significantly weaker as C_{S1}^2/C_{A1}^2 increases above unity.

REFERENCES

1. DeHoffmann, F. and E. Teller, Phys. Rev., 80, 692 (1950).
2. Shercliff, J.A., J. Fluid Mech., 9, 481 (1960).
3. Anderson, J.E., Magnetohydrodynamic Shocks, M.I.T. Monograph, Cambridge, MA (1962).
4. Kantrowitz, A.R. and H.E. Petschek, MHD characteristics and shock waves, p. 148 in Plasma Physics in theory and Application, ed. by W.B. Kunkel, McGraw-Hill, NY (1966).
5. Lax, P.D., Commun. Pure Appl. Math., 10, 537 (1957).
6. Akhiezer, A.I., G. Ia Liubarskii, and R.V. Polovin, Sov. Phys. JETP, 35, 507 (1959).
7. Jeffrey, A. and T. Tanuiti, Nonlinear Wave Propagation, Academic Press, NY (1964).
8. Brio, M. and C.C. Wu, Characteristic fields for the equations of magnetohydrodynamics, p. 19, in Nonstrictly Hyperbolic Conservation Laws, ed. by B. Keyfitz and H.C. Kranzer, American Mathematical Society, Providence, RI (1987).
9. Wu, C.C., Geophys. Res. Letts., 14, 668 (1987).
10. Wu, C.C., J. Geophys. Res., 93, 987 (1988a).
11. Wu, C.C., J. Geophys. Res., 83, 3969 (1988b).
12. Kennel, C.F., R.D. Blandford, and P. Coppi, J. Plasma Phys., 42, 299 (1989).
13. Wu, C.C., J. Geophys. Res., 95, 8149 (1990).
14. Wu, C.C. and C.F. Kennel, Phys. Rev. Letts., 68, 56 (1992).

GRAPHICAL SOLUTION OF MHD R-H RELATIONS

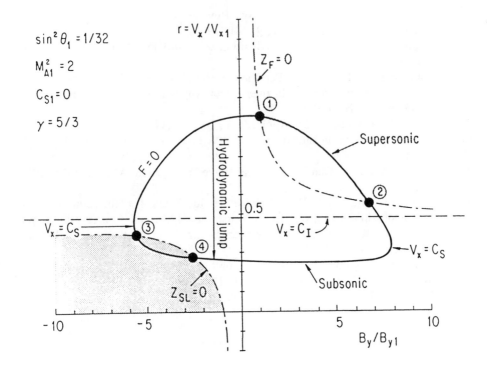

Fig. 1 Figure 1 shows the curves $F(r,b) = 0$ and $Z(r,b) = 0$ for a choice of flow parameters for which there exist four Rankine-Hugoniot states. The upper supersonic and lower subsonic branches of $F = 0$ are drawn with heavy solid lines, and the two $Z = 0$ hyperbolae are drawn with dot-dash lines. The regions where Z is positive are shaded. The dashed horizontal line is the asymptote of the $Z = 0$ hyperbolae on which $X = 0$. Since the magnitude of the local flow speed, $|V_x|$ equals the intermediate speed on $X = 0$, $|V_x|$ exceeds the local intermediate speed on the upper hyperbola, and is less than the intermediate speed on the lower hyperbola.

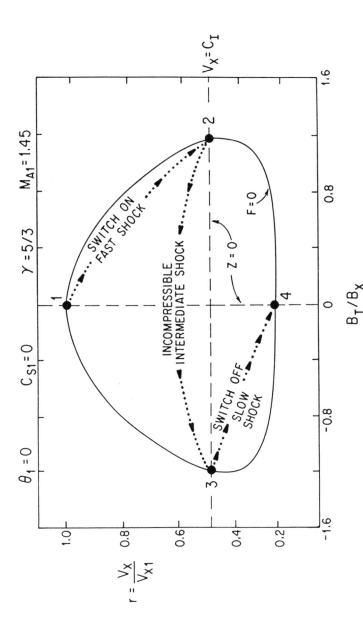

Fig. 2 This figure shows a two-dimensional planar cut through the r-axis for a case where the intermediate speed plane X = 0 does intersect the F = 0 surface. This intersection generates a circle in the X = 0 plane, which we call the 2-3 circle. The intersection of the r-axis with F = 0 generates isolated points of types 1 and 4 related by H(r) = 0. The fast switch-on shock is a transition between point 1 and any point on the 2-3 circle. The rotational discontinuity is a transition between any two points on the 2-3 circle, and the slow switch-off shock is a transition between any point on the 2-3 circle and point 4.

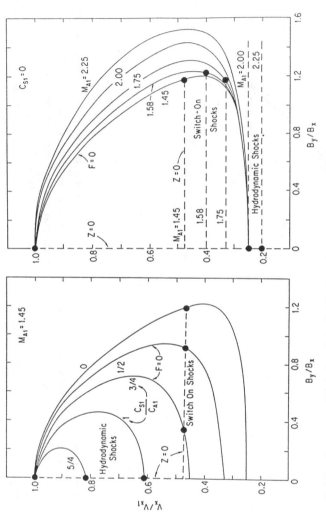

Fig. 3 The two panels plot the curves $F(r, b_y) = 0$ (solid line) and $Z(r, b_y) = 0$ (dashed line) for cases in which $M_{A1} = M_{l1}$ exceeds unity and $C_{S1} = 0$ upstream. Because of azimuthal symmetry, these curves are the same for any orientation of $\mathbf{b}_T$. When $\gamma = 5/3$, the switch-on upper limit Mach number, which is given by $M_{A1}^2 = (\gamma + 1)/(\gamma - 1)$ when $C_{S1} = 0$, is 2. Several $F = 0$ and $Z = 0$ curves are displayed in the right-hand panel for values of M_{A1} both above and below 2. The $X = 0$ plane cuts the $F = 0$ surfaces only for Alfven Mach numbers less than 2. All the $F = 0$ surfaces cross the r-axis at the same two points, which satisfy the hydrodynamic Rankine-Hugoniot relation for a strong shock, and are located at $r = 1$ and $r = (\gamma - 1)/(\gamma + 1) = 1/4$. The left-hand panel shows the same two curves for an Alfven Mach number of 1.45, and shows values of C_{S1}^2 / C_{A21}^2 for which the shock speed is both above and below the switch-on critical speed.

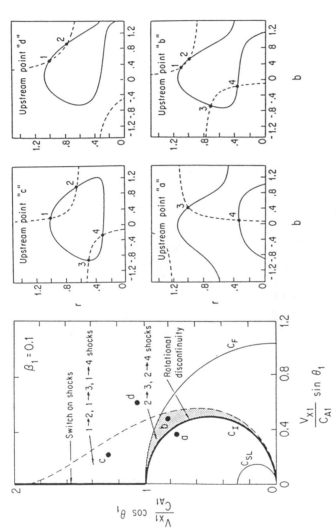

Fig. 4   The dashed line in the left-hand panel shows, for $\beta_1 = 0.1$, where the discriminant of the Rankine-Hugoniot cubic is zero in a Friedrichs diagram format. The flow parameters are defined in the normal incidence frame of the upstream state, $r = 1$, which may be a point of type 1, 2, or 3. The upstream flow speed is the radius vector and the shock propagation angle, θ_1, is the polar angle in polar coordinates. The speeds of the small amplitude fast, intermediate, and slow waves are displayed for reference. The upstream flow speeds for which there are four real Rankine-Hugoniot solutions are shaded, lightly for upstream states of type 1, and heavily for those of type 2. The dark solid line shows the speeds of switch-on shocks and the rotational discontinuity. The four panels to the right display the curves $F(r,by) = 0$ (solid lines) and $Z(r,by) = 0$ (dotted lines) that correspond to the upstream points labelled "a", "b", "c", and "d" in the Friedrichs diagram to the left.

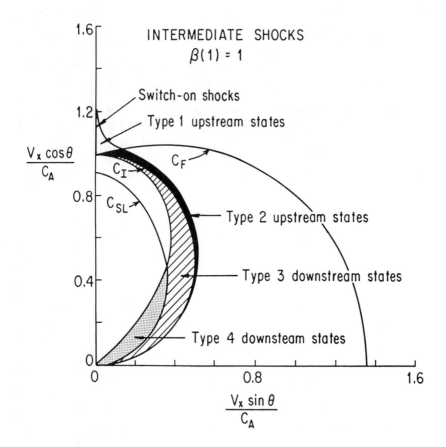

Fig. 5 This is a Friedrichs diagram for ß unity in the upstream state. Both upstream and downstream states in the region of four real Rankine-Hugoniot solutions are shown, in contrast with Figure 4. Since the switch-on critical Mach number is small when ß is unity, the region of type1 intermediate shock states is also small. Solid shading locates the type 2 states available to intermediate shocks. The hatched and dotted regions show the types 3 and 4 states available as downstream states for intermediate shocks.

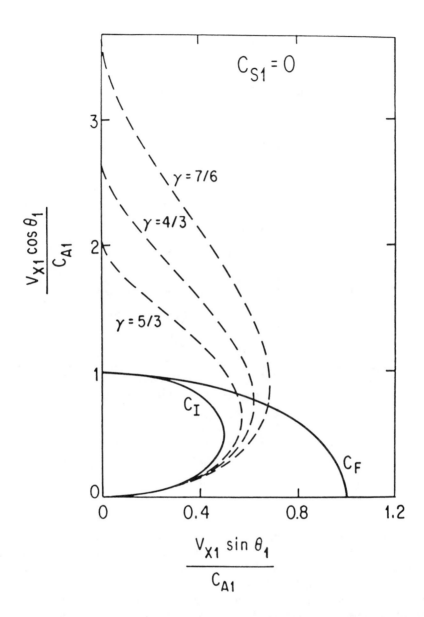

Fig. 6   This figure shows the dependence of the intermediate shock upper limit speed upon the ratio of specific heats, γ, for $C_{S1} = 0$. The switch-on critical Mach number, $M_{A1}^2 = (\gamma + 1)/(\gamma - 1)$, increases as γ decreases. Since intermediate shocks may be thought of as switch-on shocks with broken azimuthal symmetry, the upper limit speed for all intermediate shocks also increases.

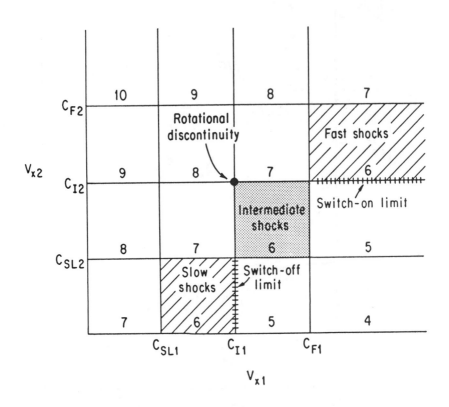

Fig. 7 The horizontal and vertical axes are the upstream and the downstream normal components of the flow speeds. Each axis is divided into four regions according to the relationship between V_x and the MHD characteristic speeds; each of the subdivisionsb corresponds to one of the Rankine-Hugoniot states, 1, 2, 3, or 4. The MHD Rankine-Hugoniot relations describe transitions between these states; the transitions between states of the same types are trivial. There remain twelve non-trivial transitions; requiring the specific entropy to increase from upstream to downstream leaves the six MHD shocks. The number of characteristics leaving each MHD transition is indicated in the appropriate box; those which satisfy the MHD evolutionary conditions are shaded. The rotational discontinuity, and the switch-on and switch-off shocks are indicated.

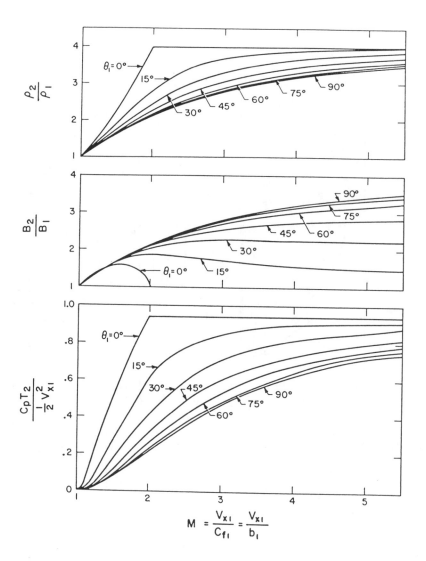

Fig. A.1 The dependences upon the fast Mach number, M, of the density compression ratio (top panel), the magnetic field compression ratio (middle panel) and the ratio of the downstream internal energy density to the upstream flow energy density (bottom panel) are shown for shock normal angles, θ, ranging in fifteen degree increments between parallel ($0°$) and perpendicular ($90°$) propagation, for $C_{S1} = 0$. (The Rankine-Hugoniot relations were solved in the standard normal incidence frame for this and all subsequent figures in this Appendix.)

MHD RANKINE-HUGONIOT SOLUTIONS

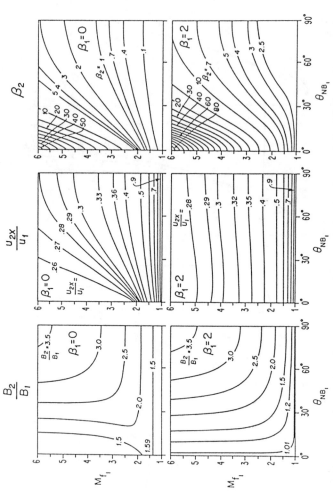

Fig. A2 This figure illustrates how fast shocks depend upon their Mach number (M_{f1}), and upon the upstream field direction (θ_{NB1}) and strength (β_1). The top and bottom rows show cases with $\beta_1 = 0$ and 2 respectively. The first column displays contours of the ratio of the downstream to upstream total magnetic field against fast Mach number and shock normal angle; the second, of the velocity contrast (here: u_{2x}/u_1); and the third, of the ratio of fluid to magnetic pressure downstream, β_2.

PARALLEL MHD FAST SHOCKS

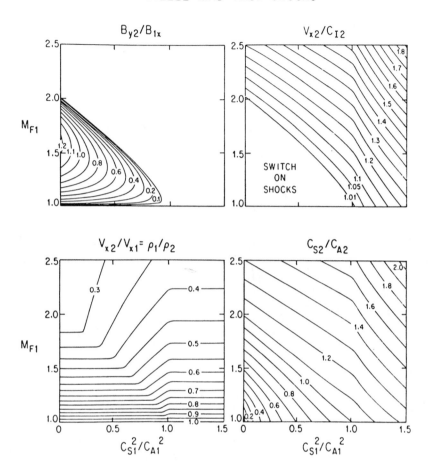

<u>Fig.A.3</u> This figure shows the dependence on C_{S1}^2/C_{A1}^2 of several properties of parallel fast shocks. Here, we define the fast Mach number, M_{F1}, to be V_{x1}/C_{A1} for $C_{S1} < C_{A1}$, and V_{x1}/C_{S1} for $C_{S1} > C_{A1}$, and contour the ratios of the downstream transverse magnetic field to the upstream total field (upper left panel), of the normal component of the downstream flow speed to the intermediate speed (upper right), and of the sound speed to the Alfven speed downstream (lower right), together with the denstiy compression ratio (lower left), against C_{S1}^2/C_{A1}^2 and M_{F1}. The switch-on shock region is where B_{y2} is nonzero and the downstream flow speed equals the intermediate speed. The 1-2 hydrodynamic shock is in the complement of the switch-on region.

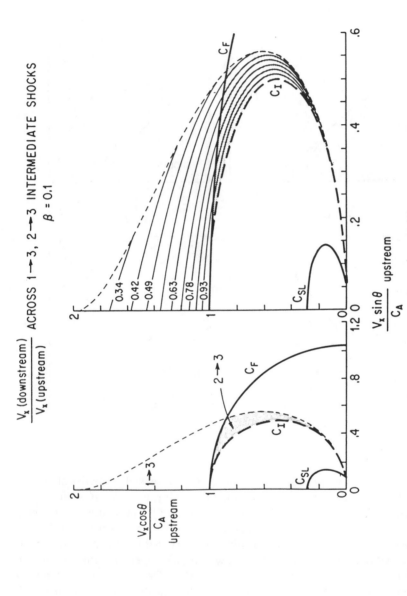

Fig. A.4 This figure contours the velocity contrast of 1-3 and 2-3 intermediate shocks in a Friedrichs diagram format for an upstream $\beta = 0.1$. The panel to the left of the figure shows the upper limit speed for intermediate shocks (dashed line) and the right-hand panel displays the velocity contrasts. We have exaggerated the horizontal scales in the right-hand panel to emphasize the intermediate shock regions.

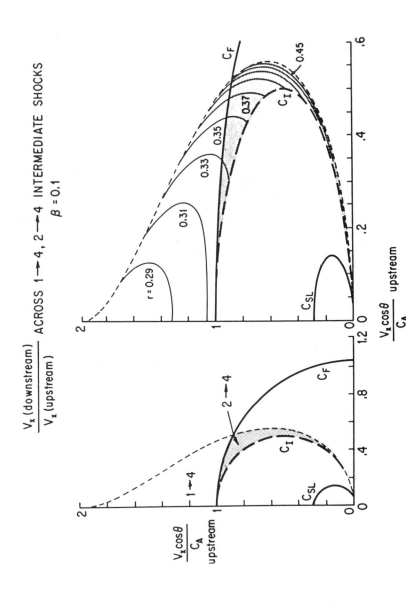

Fig. A.5 This figure contours the velocity contrasts of 1-4 and 2-4 intermediate shocks in a Friedrichs diagram format for an upstream β = 0.1. The panel to the left of the figure shows the upper limit speed for intermediate shocks (dashed line) and the right-hand panel displays the velocity contrasts. We have exaggerated the horizontal scales in the right-hand panels to emphasize the intermediate shock regions.

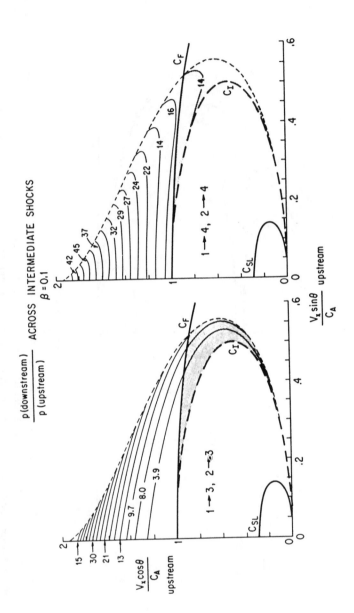

Fig. A.6 This figure contours the ratio of the downstream to upstream pressure for the 1-3, 1-4 (left) and 2-3, 2-4 (right) intermediate shock families. Once again, the 1-3, 2-3, shocks are weak when V_{x1} is close to the intermediate speed, while the 2-3, 2-4 shocks never have small pressure ratios. In either case, the pressure ratios are large for shock speeds near the upper limit speed. Thus, strong intermediate shocks can heat the plasma significantly.

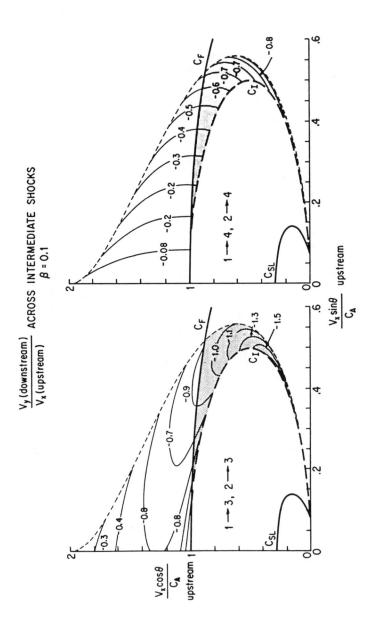

<u>Fig. A.7</u> These figures contour the ratios of the downstream transverse velocity to the upstream normal velocity for the 1-3, 2-3 (left panel) and the 1-4, 2-4 (right panel) intermediate shocks.

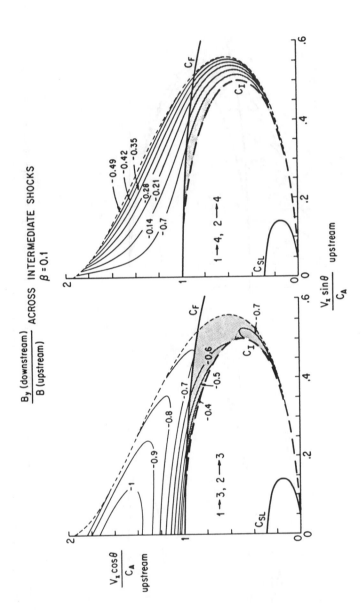

Fig. A.8 These figures contour the ratios of the downstream transverse magnetic field to the upstream total field for the 1-3, 2-3 (left panel) and the 1-4, 2-4 (right panel) intermediate shocks.

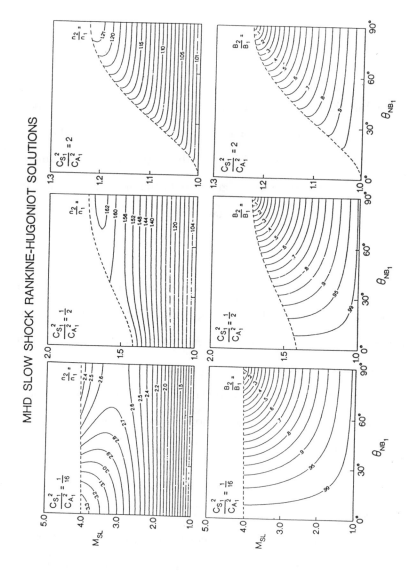

Fig. A.9 This figure contours the density compression ratio (here: n2/n1) and total magnetic field compression ratio against the slow Mach number, MSL, and shock normal angle (θNB1) for $C_{S1}^2/C_{A1}^2 = 1/16$ (left column), 1/2 (middle), and 2 (right). The switch-on Mach number, MSL = C_{I1}/C_{SL1}, is indicated by a dashed line in each panel.

DRIFT PLASMA DISPERSION FUNCTION

S. T. Tsai, S. C. Guo, M. Yu

Institute of Physics, P. O. Box 603, Beijing 100080 China

ABSTRACT

The analytic properties of the drift plasma dispersion function are studied in detail via extending the bivariated normal probability function to the complex plane with analytic continuation. The recurrence relations are derived. Some limiting cases are discussed. It is useful for studying the collisionless kinetic responses, e.g., instabilities driven by magnetic field inhomogeneities, and/or weak relativistic effects in anisotropic plasmas.

I. INTRODUCTION

Recently, the kinetic emission and absorption of the energetic and/or high β inhomogeneous plasmas have attracted a great deal of attention in both fusion and space studies. These efforts have been spurred on by urgent research programs of current interest such as instabilities driven by energetic particles, magnetic field gradients and curvature, cyclotron maser and heating etc. A general local dispersion relation retaining both inhomogeneity and weak relativistic effects has been derived[1].

In order to study the analytic properties of the fluctuations, it is necessary to develop a drift plasma dispersion function (DPDF). Similar studies have been given by Shkarofsky[2] and others[3-6] for weak relativistic isotropic cyclotron heating (using F function) by Similon et al.[7] (numerically), by Tsai et al.[8] (using expansion method), and Kim et al.[9] (via specific modeling). In this paper we extend the bivariated normal probability function used in statistical mathematics to the complex plane. The recurrence relations, analytic continuation, and the resonance pole properties are studied in detail.

II. DRIFT PLASMA DISPERSION FUNCTION

It is well known that for a plane wave $\exp(ik \cdot x - i\omega t)$, the Landau pole integration leads to the plasma dispersion function $Z[(\omega - l\omega_c)/(k_\| v_\|)]$. For studying the local magnetic inhomogeneous and/or weak relativistic kinetic resonance effects derived from gyrokinetic equations,[1,10] one encounters the velocity integration:

$$W_{m,n} \equiv 2\pi \int_0^\infty \int_{-\infty}^\infty u\,du\,dv \frac{u^{2(m-1)} v^{n-1} F(u^2, v)}{\Omega + au^2 + bv^2 - 2K_\| v}, \qquad (1)$$

where $F(u^2, v)$ is local plasma distribution function and a displaced bi-Maxwellian:

$$F(u^2, v) = \frac{1}{\pi^{3/2}} \exp[-u^2 - (v - V)^2] \qquad (2)$$

will be considered; u, v are dimensionless relativistic velocities perpendicular and parallel to the B field respectively, and they are implicitly spatially dependent via normalization factors $u = v_\perp/\alpha_\perp$, $v = v_\|/\alpha_\|$; m and n are positive integers;

$$\Omega = \omega' - l\omega_c, \quad \omega' = \omega - \omega_E,$$

$$a = \frac{\alpha_\perp^2}{2c^2}\omega' - \omega_{\nabla B} - \frac{\alpha_\perp^2 k_\|}{2\omega_c}R_1, \quad b = \frac{\alpha_\|^2}{2c^2}\omega' - 2\omega_k,$$

$$2K_\| = \alpha_\|(k_\| - lR_2), \quad R_1 = \vec{e}_\perp \cdot \vec{\nabla}_x \times \vec{e}_\|,$$

$$R_2 = \vec{e}_1 \cdot (\vec{e}_\| \cdot \vec{\nabla}_x \vec{e}_2) - \frac{\vec{e}_\| \cdot (\vec{\nabla}_x \times \vec{e}_\|)}{2}, \quad V = \frac{V_0}{\alpha_\|};$$

$\alpha_\perp$, $\alpha_\|$ are the perpendicular and parallel thermal velocity; V_0 is the parallel mean velocity, ω_c, ω_E, $\omega_{\nabla B}$ and ω_κ are the gyro, $\vec{E} \times \vec{B}$, ∇B and curvature B frequencies; and l is an integer. Ω, a, b can be complex via ω. The causality condition requires $\omega_i > 0$ and this gives rise to $\text{Im}(\Omega, a, b) > 0$.

Since the denominator of the integrand is quadratic in u and v, $W_{m,n}$ can be expressed by a linear combination of $W_{m,1}$ and $W_{m,2}$. Using the relation

$$(\Omega + au^2 + bv^2 - 2K_\parallel v)^{-1} = -i \int_0^\infty dt \exp[i(\Omega + au^2 + bv^2 - 2K_\parallel v)t],$$

and, after some algebra, we have

$$\begin{pmatrix} W_{m,1} \\ W_{m,2} \end{pmatrix} = \Gamma(m) \begin{pmatrix} D_{m,1} \\ -yD_{m,2} + \frac{K_\parallel}{b} D_{m,1} \end{pmatrix}, \tag{3}$$

where DPDF has been defined as:

$$D_{m,n}(z, a, b, y) \equiv -i \int_0^\infty dt \frac{\exp[izt + y^2/(1 - ibt) - y^2]}{(1 - iat)^m (1 - ibt)^{n-1/2}}, \tag{4}$$

$$z \equiv \Omega - \frac{K_\parallel^2}{b}, \quad y = \frac{K_\parallel}{b} - V, \quad z_i = \omega_i \left(1 + \frac{K_\parallel^2}{|b|^2} \frac{\alpha_\parallel^2}{2c^2}\right).$$

By the definition of $D_{m,n}$ in Eq.(4), the following recurrence relations can be obtained:

$$a\frac{d}{dz}D_{m,n} = D_{m,n} - D_{m-1,n} \tag{5}$$

$$b\frac{d}{dz}D_{m,n+1} = D_{m,n+1} - D_{m,n} \tag{6}$$

$$\frac{d}{dy^2}D_{m,n} + D_{m,n} = D_{m,n+1} \tag{7}$$

$$(a - b)D_{m,n} = aD_{m,n-1} - bD_{m-1,n} \tag{8}$$

$$zD_{m,n} + maD_{m+1,n} + (n - \frac{1}{2})bD_{m,n+1} + y^2bD_{m,n+2} = 1 \tag{9}$$

$$\left(a\frac{d}{da} + b\frac{d}{db} + \frac{d}{dz}z\right)D_{m,n} = 0 \tag{10}$$

$$\frac{d}{da}D_{m,n} = m\frac{d}{dz}D_{m+1,n} \tag{11}$$

$$\frac{d}{db}D_{m,n} = (n - \frac{1}{2})\frac{d}{dz}D_{m,n+1} + y^2\frac{d}{dz}D_{m,n+2} \tag{12}$$

Function $D_{m,n}$ can be generated from Eqs. (7)—(9), if $D_{1,1}$ and $D_{0,1}$ are given. By Eq. (4), we know

$$D_{0,n}(z, a, b, y) = D_{1,n}(z, a = 0, b, y).$$

And from Eq. (16), we get

$$D_{0,1} = \frac{1}{2b\sqrt{-z/b}} \left\{ Z \left[\sqrt{-\frac{z}{b}} - y \right] + Z \left[\sqrt{-\frac{z}{b}} + y \right] \right\}, \tag{13}$$

where Z is the plasma dispersion function. It is worth mentioning that if $y = 0$, the $D_{0,n}$ function is reduced to the $2c^2/(\alpha_\parallel^2 \omega) F_{n-1/2}$ function of Shkarofsky[2]. Therefore, the only function needed to be evaluated analytically in great detail is $D_{1,1}$. From Eq. (3) we know

$$W_{1,1} = D_{1,1}. \tag{14}$$

III. THE ANALYTIC PROPERTIES OF $D_{1,1}$

By Eqs. (1), (2) and (3) we have

$$D_{1,1} = W_{1,1} = \frac{1}{\sqrt{\pi}} \int_0^\infty d\xi \int_{-\infty}^\infty dv \frac{e^{-\xi-(v-V_0)^2}}{\Omega + a\xi + bv^2 - 2K_\parallel v} \tag{15}$$

$$= \int_0^\infty e^{-\xi} d\xi \left\{ \frac{1}{2b\sqrt{-z'/b}} \left[Z \left(\sqrt{-\frac{z'}{b}} - y \right) + Z \left(\sqrt{-\frac{z'}{b}} + y \right) \right] \right\} \tag{16}$$

where $\xi \equiv u^2$, $z' = a\xi + z$.

Taking the branch cut along the positive real axis on the $-z'/b$ plane, the causality $\omega_i > 0$ gives

$$\text{Im} \sqrt{-\frac{z'}{b}} > \text{Im } y \geq 0 \tag{17}$$

for positive ξ. Let $x = \sqrt{-(a\xi + z)/b}$, we have

$$D_{1,1} = \sum_{\pm} \int_0^\infty e^{-\xi} d\xi \frac{2i}{2b\sqrt{-(a\xi + z)/b}} \exp\left[-\left(\sqrt{-\frac{a\xi + z}{b}} \pm y\right)^2 \right]$$

$$\times \int_{-\infty}^{i\sqrt{2}[\sqrt{-(a\xi+z)/b}\pm y]} e^{-t^2/2} \frac{dt}{\sqrt{2}}$$

$$= \sum_{\pm} \int_{\sqrt{-z/b}}^{\sqrt{-(a\infty+z)/b}} dx \exp(\frac{a}{b}x^2 + \frac{z}{a})(-\frac{\sqrt{2}i}{a})e^{-(x\pm y)^2} \int_{-\infty}^{i\sqrt{2}(x\pm y)} e^{-t^2/2} dt.$$

Then we let $\eta = \sqrt{2}\sqrt{1-b/a}[x \pm y/(1-b/a)]$ (here we take branch cut of $\sqrt{1-b/a}$ to be on the positive real axis). After some algebra we obtain

$$D_{1,1} = -\frac{i}{a}\frac{2\pi}{\sqrt{1-b/a}} \exp\left(\frac{z}{a} + \frac{b}{a}\frac{y^2}{1-b/a}\right)$$

$$\sum_{\pm} \frac{1}{2\pi} \int_{\eta_{0\pm}}^{\eta_{\infty}} e^{-\eta^2/2} d\eta \int_{(\kappa_{\pm}-\rho\eta)/\sqrt{1-\rho^2}}^{\infty} e^{-t^2/2} dt$$

$$= -\frac{i}{a}\frac{2\pi}{\sqrt{1-b/a}} \exp\left(\frac{z}{a} + \frac{b}{a}\frac{y^2}{1-b/a}\right) \sum_{\pm} L_{\pm}(\eta_{\infty}, \eta_{0\pm}, \kappa_{\pm}, \rho), \quad (18)$$

where

$$\eta_{0\pm} \equiv \sqrt{2}\sqrt{1-\frac{b}{a}}\left[\sqrt{-\frac{z}{b}} \pm \frac{y}{1-b/a}\right]$$

$$\eta_{\infty} \equiv \sqrt{2}\sqrt{1-\frac{b}{a}}\sqrt{-\frac{a}{b}\infty}$$

$$\rho \equiv -\sqrt{\frac{a}{b}}$$

$$\kappa_{\pm} \equiv \mp i\frac{\sqrt{2}y}{\sqrt{1-a/b}},$$

and

$$L(\eta_{\infty}, \eta_0, \kappa, \rho) = \frac{1}{2\pi} \int_{\eta_0}^{\eta_{\infty}} d\eta \int_{(\kappa-\rho\eta)/\sqrt{1-\rho^2}}^{\infty} dt e^{-(\eta^2+t^2)/2} \quad (19)$$

is the bivariated normal probability function defined on the complex plane due to complex ω. Here L is a convergent function.

In the complex plane, one can prove that

$$L(\eta_{\infty}, \eta_0, \kappa, \rho) = L(\eta_{\infty}, 0, 0, \rho) - \frac{1}{4}A(\eta_0) + V(\eta_0, \frac{\kappa/\eta_0 - \rho}{\sqrt{1-\rho^2}})$$

$$-\frac{1}{4}A(\kappa) + V(\kappa, \frac{\eta_0/\kappa - \rho}{\sqrt{1-\rho^2}}), \quad (20)$$

where

$$A(\kappa) = \frac{1}{\sqrt{2\pi}} \int_{-\kappa}^{\kappa} e^{-\tau^2/2} d\tau$$

$$V(h,\kappa) = \frac{1}{\sqrt{2\pi}} \int_0^h d\tau \int_0^{\kappa\tau} d\eta\, e^{-(\tau^2+\eta^2)/2}.$$

After a fair amount of algebraic manipulations we finally obtain

$$D_{1,1} = \frac{i2\pi}{a\sqrt{1-b/a}} \exp\left(\frac{z}{a} + \frac{b}{a}\frac{y^2}{1-b/a}\right) \left\{ i\frac{1}{\pi} \ln\left[-i\sqrt{-\frac{a}{b}} + i\sqrt{-\frac{a}{b}}\sqrt{1-\frac{b}{a}}\right] \right.$$

$$+ \frac{1}{4}\sum_{\pm} A\left[\sqrt{2}\left(\sqrt{-\frac{z}{b}}\sqrt{1-\frac{b}{a}} \pm \frac{y}{\sqrt{1-b/a}}\right)\right]$$

$$\left. - \sum_{\pm} V\left[\sqrt{2}\left(\sqrt{-\frac{z}{b}}\sqrt{1-\frac{b}{a}} \pm \frac{y}{\sqrt{1-b/a}}\right), -i\frac{\sqrt{-z/b}\pm y}{\sqrt{-z/b}\sqrt{1-b/a}\pm \frac{y}{\sqrt{1-b/a}}}\right] \right\}$$

$$\equiv \mu\left\{\frac{i}{\pi}\ln\alpha + \frac{1}{4}\sum_{\pm} A(\beta_\pm) - \sum_{\pm} V\left(\beta_\pm, -i\sqrt{2}\frac{\nu_\pm}{\beta_\pm}\right)\right\} \qquad (21)$$

where we have used $A(-\kappa) = -A(\kappa)$, $V(\kappa_+, \frac{\eta_{0+}/\kappa_+ -\rho}{\sqrt{1-\rho^2}}) = -V(\kappa_-, \frac{\eta_{0-}/\kappa_- -\rho}{\sqrt{1-\rho^2}})$ and introduced abbreviated notations: μ, α, $\beta_\pm$ and $\nu_\pm$. One notices that $D_{1,1}$ is an analytic function. Also, ln, A, V are analytic functions for a and $b \neq 0$.

As in the previous derivations, taking the branch cuts of $\sqrt{-z/b}$, $\sqrt{-a/b}$ and $\sqrt{1-b/a}$ along positive real axes and $\ln\alpha$ along the negative real axis, one can prove that for $\omega_i > 0$, $-a/b$ and $-z/b$ can not be positive real value. When ω_i switches sign, $\sqrt{-z/b}$, $\sqrt{-a/b}$ and $\sqrt{1-b/a}$ go to the next Riemann sheet only at $a_r b_r < 0$ or $z_r b_r < 0$. For real a, b and z, we treat them as the limiting cases of $\omega_i = 0^+$. One can also prove α can not be a negative real value, so that $\ln\alpha$ always stays in one Riemann sheet. To guarantee the analytic continuation from the upper half ω-plane ($\omega_i > 0$) to the entire ω-plane (including $\omega_i \leq 0$), the DPDF can be rewritten as

$$D_{1,1} = \mu\left[\frac{i}{\pi}\ln(S_1\alpha) + \frac{S_2}{4}\sum_{\pm} A(\beta_\pm) - \sum_{\pm} V\left(\beta_\pm, -i\sqrt{2}\frac{\nu_\pm}{\beta_\pm}\right)\right], \qquad (22)$$

where

$$S_1 = \begin{cases} -1 & \text{for } \omega_i < 0 \text{ and } a_r b_r < 0 \\ \text{sign}(-b) & \text{for } \omega_i = 0 \text{ and } ab < 0 \\ 1 & \text{otherwise} \end{cases} \tag{23}$$

$$S_2 = \begin{cases} -1 & \text{for } \omega_i < 0 \text{ and } z_r b_r < 0 \\ \text{sign}(-b) & \text{for } \omega_i = 0 \text{ and } zb < 0 \\ 1 & \text{otherwise.} \end{cases} \tag{24}$$

In the above derivation, we have used the properties of $A(S_2\beta_\pm) = S_2 A(\beta_\pm)$, $V(-h,\kappa) = V(h,\kappa)$ and $D_{1,1}$ even in y. The singular cases, $a = 0$, $b = 0$ and $a = b$, which have been overlooked in the above discussion, will be studied in the next section.

IV. SOME LIMITING CASES

A. For $\omega_i = 0$ (i.e. a, b, z are real), the following result is deduced:

(1) $\mu \frac{i}{\pi} \ln(S_1 \alpha)$ is real for $b/a > 0$ and becomes complex if $b/a < 0$,

(2) $\mu \frac{S_2}{4} \sum_\pm A_\pm(\beta_\pm)$ is real for $z/b > 0$ and imaginary if $z/b < 0$,

(3) $\mu \sum_\pm V_\pm(\beta_\pm, -i\sqrt{2}\nu_\pm/\beta_\pm)$ is real for all z/a and b/a,

where we have used the relations $A(Z^*) = A^*(Z)$ and $V(h^*, a^*) = V^*(h, a)$. Therefore, at the marginal stability the first and/or the second term of $D_{1,1}$ in Eq. (22) can be nonreal depending on the signs of a, b and z.

For $\text{Im}(a, b) = 0$ and $z_i \ll z_r$: the same properties as above are obtained for the first two terms of Eq. (22). However, the third term becomes complex.

B. For small a case, by rearranging the terms in $\beta_\pm$ and carrying out part of the integration we obtain

$$D_{1,1} = \frac{1}{2}\exp(\frac{z}{b} - y^2)\sum_\pm \left\{ \frac{\exp(\mp 2y\sqrt{-z/b})}{(a-b)\sqrt{-z/b} \pm ay} \left[\sqrt{\pi}\beta_\pm Z(-iS_2\bar{\beta}_\pm) \right.\right.$$
$$\left.\left. + \sum_{n=0}^{\infty} \frac{1}{\bar{\beta}_\pm^{2n}} \gamma(n + \frac{1}{2}, \nu_\pm^2) \right] \right\}, \tag{25}$$

where

$$\bar{\beta}_\pm \equiv S_3 \sqrt{\frac{z}{a}}\sqrt{1 - \frac{a}{b}} \pm S_1 y \frac{\sqrt{-a/b}}{\sqrt{1-a/b}},$$

$$S_3 = \begin{cases} -1 & \text{for } \omega_i < 0 \text{ and } a_r z_r < 0 \\ & \text{or } \omega_i = 0 \;,\; a < 0 \text{ and } z > 0 \\ 1 & \text{otherwise} \end{cases} \tag{26}$$

and $\gamma(n + 1/2, \nu_\pm^2) = \int_0^{\nu_\pm^2} e^{-x} x^{n-1/2} dx$ is the incomplete Γ-function.

For $a = 0$ one can recover Eq. (13).

C. For small $(1 - b/a)$ case, by the similar method used in case B, one can get

$$D_{1,1} = \frac{1}{2a\sqrt{1 - b/a}} \exp(\frac{z}{b} - y^2) \sum_\pm \left\{ \exp\left(\mp 2y\sqrt{-\frac{z}{b}}\right) \left[\sqrt{\pi} Z(-iS_2 \frac{\beta_\pm}{\sqrt{2}}) \right. \right.$$

$$\left. \left. + \sum_{n=0}^{\infty} \frac{2^{n+1/2}}{\beta_\pm^{2n+1}} \gamma(n + \frac{1}{2}, \nu_\pm^2) \right] \right\}, \tag{27}$$

where we have used $S_1 = 1$ when $|1 - b/a| \ll 1$. If $a = b$ by the definition of Eq. (4) we have $D_{1,1}(a = b) = D_{0,2}$, or, more generally,

$$D_{m,n}(a = b) = D_{0,m+n}. \tag{28}$$

D. For small b ($b \ll a$, $K_\parallel^2/\Omega$, $K_\parallel/V$) case, we expand Eq. (22) and work out the integrations. Two divergent terms in V function can be cancelled by the other two divergient terms in $\ln$ and A functions. The final expression of $D_{1,1}(b \simeq 0)$ is very complicated. Only the lowest order terms are presented here.

$$D_{1,1} = -\frac{i2\pi}{a} \exp[-f_1] \left\{ -\frac{i}{\pi} \left[\ln(-S_1 S_4) + \ln\frac{K_\parallel}{a}\right] \right.$$

$$+ \frac{S_4}{4} A(f_2) + V\left(f_2, i\frac{\Omega/K_\parallel - 2V}{\sqrt{2} f_2}\right) - \frac{1}{4} - \frac{i}{2\pi} \ln 2 \tag{29}$$

$$\left. + \frac{i}{4\pi} [E_1(f_1) + \ln(f_1)] \right\} + O(b),$$

where

$$f_1 \equiv -K_\parallel^2 (1/a + \Omega/K_\parallel^2 - 2V/K_\parallel)/a, \quad f_2 \equiv K_\parallel(2V/K_\parallel - \Omega/K_\parallel^2 - 2/a)/\sqrt{2},$$

$S_4 = S_2 \, \text{sign}(b_r)$, and E_1 is the exponential integral function. The higher order terms can be derived straightforwardly.

If both b and $K_{\parallel}$ vanish, Eq.(15) gives

$$D_{1,1}(b = K_{\parallel} = 0) = \int_0^{\infty} d\xi \frac{e^{-\xi}}{a\xi + \Omega} = -\frac{\exp(\Omega/a)}{a} E_i(-\frac{\Omega}{a}). \qquad (30)$$

From the above limiting cases B-D, one can find that Eq. (22) is also valid for the cases $a = 0$, $b = 0$ or $a = b$ if the branch cuts are carefully considered.

V. CONCLUSION

In this paper we have defined DPDF and expressed it in terms of the complex bivariated normal probability function and the Z function. The expression is simplified by considering that $D_{1,1}$ is even in y and by using the analytic properties of the V and A functions. The analytic continuation of $D_{1,1}$ is realized by taking proper values of S_1 and S_2 in Eq. (22). Although only $D_{1,1}$ and $D_{0,1}$ have explicity been given in the paper, the recurrence relations can be adopted to generate any $D_{m,n}$. The simpler results are obtained for the limiting cases of small a, $1 - b/a$ and b and presented in Eqs. (25), (27) and (29). These analytic expressions of the DPDF are useful in studying the high β or energetic inhomogeneous plasmas, which will be the subject of forthcoming papers.

ACKNOWLEDGEMENTS

The authors gratefully acknowledge valuable discussions with Prof. T. H. Stix, Dr. J. W. Shen, and Dr. H. D. Xu. This work was supported by NSFC, AAAPT and Grant No. LWTZ-1298 of CAS.

REFERENCES

1. S. C. Guo and S. T. Tsai, ACTA Physica Sinica **36**, 870 (1987).

2. I. P. Shkarofsky, Phys. Fluids **9**, 561 (1966).

3. S. T. Tsai, C. S. Wu, Y. D. Wang, and S. K. Kang, Phys. Fluids **24**, 2186 (1981).

4. A. C. Airoldi and A. Orefice, J. Plasma Phys. **27**, 515 (1982).

5. V. Krivenski and A. Orefice, J. Plasma Phys. **30**, 125 (1983).

6. M. Bornatici, R. Cano, O. DeBarbieri, and F. Engelmann, Nuclear Fusion **23**, 1153 (1983).

7. P. Similon, J. E. Sedlak, D. Stotler, H. L. Berk, W. Horton, and D-I. Choi, J. Comp. Phys. **54**, 260 (1984).

8. S. T. Tsai and S. C. Guo, In"Sino-Japan Bilateral Workshop on Statistical and Condensed Matter Theory", ed. by Xie Xide (World Scientific Pub. Co., Singapore, 1986) p. 240

9. J. Y. Kim, Y. Kishimoto, W. Horton, and T. Tajima, IFSR623 (1993).

10. S. T. Tsai, J. W. Van Dam, and L. Chen, Plasma Phys. **26**, 907 (1984).

Author Index

B

Berk, H. L., 140
Breizman, B. N., 140

D

Davidson, R. C., 1
Dawson, J. M., 39

H

Hecht, J., 26
Hosea, J. C., 128

K

Kennel, C. F., 180
Kruer, W. L., 16

L

Lund, S. M., 1

M

Ma, S., 39
McCrory, R. L., 26

N

Nunan, W. J., 39

O

Ofer, D., 26
Ono, M., 156
Orszag, S. A., 26

P

Porkolab, M., 99

S

Shvarts, D., 26

T

Tsai, S. T., 228

W

Wilks, S. C., 16
Wong, A. Y., 69

Z

Zinamon, Z., 26